TOPICS IN
NUCLEIC ACID STRUCTURE

TOPICS IN NUCLEIC ACID STRUCTURE

Edited by

STEPHEN NEIDLE

Department of Biophysics
University of London King's College

A HALSTED PRESS BOOK

John Wiley & Sons
New York

First published in Great Britain 1981 by
Macmillan Publishers Ltd
London and Basingstoke

Published in the USA by
Halsted Press, a Division of
John Wiley & Sons, Inc, New York

Printed in Great Britain

Library of Congress Cataloging in Publication Data
Main entry under title:

Topics in nucleic acid structure.

(Topics in molecular and structural biology)
"A Halsted Press book."
Includes index.
1. Nucleic acids.   2. Molecular structure.
I. Neidle, Stephen. II. Series.
QP620. T66   574.87'328   81–1316
ISBN 0–470–27161–2   AACR2

Contents

Contents

The Contributors

S. Arnott,
Department of Biological Sciences,
Purdue University,
West Lafayette,
Indiana 47907,
USA

M. M. Dhingra,
Department of Chemistry,
State University of New York at Albany,
Albany,
New York 12222,
USA

Permanent address:
Tata Institute of Fundamental Research,
Bombay,
India

T. R. Krugh,
Department of Chemistry,
University of Rochester,
Rochester,
New York 14627,
USA

S. Neidle,
Department of Biophysics,
University of London King's College,
26-29 Drury Lane,
London WC2B 5RL,
UK

R. H. Sarma,
Department of Chemistry,
State University of New York at Albany,
Albany,
New York 12222,
USA

H. M. Berman,
The Institute for Cancer Research,
Fox Chase Cancer Center,
7701 Burholme Avenue,
Fox Chase,
Philadelphia 19111,
USA

S. H. Kim,
Department of Biochemistry,
Duke University Medical Center,
Durham,
North Carolina 27710,
USA

D. M. J. Lilley,
Research Division,
G. D. Searle and Company Ltd.,
P.O. Box 53,
Lane End Road,
High Wycombe,
Bucks HP12 5RL,
UK

B. R. Reid,
Department of Biochemistry,
University of California,
Riverside,
California 92521,
USA

H.-S. Shieh,
The Institute for Cancer Research,
Fox Chase Cancer Center,
7701 Burholme Avenue,
Fox Chase,
Philadelphia 19111,
USA

Preface

The study of nucleic acid structure has mushroomed into a subject of its own since the original suggestion for the double-helical nature of DNA was made over a quarter of a century ago, with ramifications in many aspects of molecular and structural biology. Since then, we have seen increasingly detailed physical examinations of the nucleic acids themselves, as well as of oligo- and mono-nucleotides — it has often been felt that the true path to a detailed understanding of the polymers lies via consideration of their constituent units. Much of this volume is therefore necessarily concerned with such a reductionist approach, which to a large extent has been vindicated by the accumulated structural data particularly over the past few years. The dominance of x-ray diffraction and crystallographic methods of determining molecular structure and conformation is reflected in the emphasis accorded them here. They have been increasingly complemented by the advent of ever-more-powerful magnetic resonance methods for the study of conformation in solution; this development too is given due emphasis in several chapters.

In recent years the subject has come full circle with a revival of interest in DNA structure itself. This has been stimulated by advances in biochemical and biophysical studies on the repeating unit of the genome. The fundamental questions of how the DNA is organised within these nucleosome units are currently the subject of frenzied activity in many laboratories, as well as of considerable speculation. It can only be a matter of time (and a good deal of effort) before the crystal structure of the nucleosome is obtained, which, one hopes, will go at least some way towards answering these questions.

The determination of the crystal structure of yeast phenylalanine transfer RNA represents a landmark in the subject in that it was the first true nucleic acid structure to be determined at anything approaching atomic resolution—it is perhaps a measure of the achievements of the groups involved that no other transfer RNA has as yet been so analysed. Transfer RNA has provided the NMR experts with a favourable opportunity to evaluate questions of solid state versus solution conformations.

Crystallographic analyses by their nature provide us with a static picture of nucleic acid molecules. Thus, the available conformational data available to date

ix

that is directly relevant to their biological behaviour is rather scanty; the long-held hope of the structure determination of a tRNA-synthetase complex, for example, has not yet materialised. However, the area of small-molecule-nucleic acid interactions is now beginning to yield information of this type, although there is a long path to tread before conformational changes can be related to biological properties of the drugs involved.

It is inevitable that a book of this nature becomes dated all too quickly; however it is hoped that it represents a fair and reasonably balanced account of the fundamental features of nucleic acid structure that will be of use to a wide range of readers for some time to come.

I am grateful to many friends and colleagues for the benefit of their advice and wisdom. Particular thanks go to Helen Berman, Watson Fuller, Michael Spencer, Peter Young and Maurice Wilkins, and to my wife Andrea for her help in many ways.

London, 1980   S.N.

1

Conformational principles of nucleic acids

Helen M. Berman

INTRODUCTION

From a conformational standpoint, nucleic acids are enormously complex. In order to rigorously describe, for example, the secondary structure of a repeating poly-nucleotide such as fibrous poly dA·dT, the values of twelve torsion angles must be ascertained (Figure 1.1); contrast this with a repeating polypeptide where only two conformation angles need be determined to describe its secondary structure. For a relatively small single-stranded nucleic acid with tertiary structure such as tRNA, a complete conformational description necessitates the determination of almost one thousand torsion angle values. The challenging task undertaken in nucleic acid structural studies is, therefore, to elucidate the principles which underly the folding of polynucleotides, and ultimately to be able to predict structures.

In this chapter are set forth the definitions and nomenclature used to describe nucleic acids, followed by a summary of what is known about some of the conforma-tional angles. Finally, the applications of conformational principles to the under-standing of the structures of DNA, tRNA and chromatin are briefly discussed.

DEFINITIONS AND NOMENCLATURE

The structures of polynucleotides and their constituents are described in terms of their atomic coordinates, or more commonly in terms of their conformational or torsion angles. These angles are the *projected* angles between two adjacent bonds when viewed along a central bond. Figures 1.1 and 1.2 show the conventions used throughout this book. For the polynucleotide backbone (Seeman *et al.*, 1976), the C3'—O3' bond is used as the starting point and its torsion angle is designated as α. Moving successively in the 5' to 3' direction, O3'—P is β, P—O5 is γ, O5'—C5 is δ, C5'—C4' is ϵ and C4'—C3' is ζ (Figure 1.1). The torsion angles in the ribose sugar ring (Figure 1.2a) are τ_0 for O1'—C1', τ_1 for C1'—C2', τ_2 for C2'—C3', τ_3 for C3'—C4' and τ_4 for C4'—O1' (Sundaralingam, 1969). Rotation about the glycosidic bond (χ) is defined with respect to O1' and C6 in pyrimidines and O1'

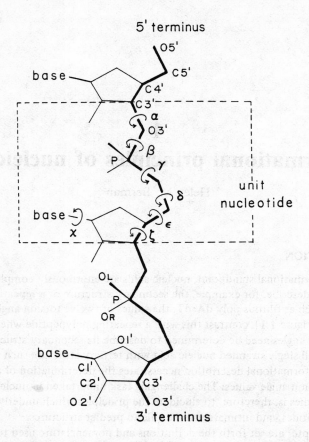

Figure 1.1   The conformational nomenclature for the nucleotide backbone. Darkened lines indicate atoms used to define torsion angles

and C8 in purines (Figure 1.2b). For all these torsion angles, values are positive for a right-hand rotation. Table 1.1 summarises some of the more popular conventions and may serve as a useful guide in surveying the conformational literature.

In addition to the descriptions of conformations in terms of torsion angles, certain qualitative descriptions are used. In the case of alternative descriptions the ones in use in this article are indicated. In the conventions of most authors *cis* denotes a folded conformation with a torsion angle of 0°*, a *trans* conformation

*Because Olson and Flory (1972) assign their 0° value to an extended or *trans*(*t*) conformation, in their scheme a *cis* conformation has a value of 180°.

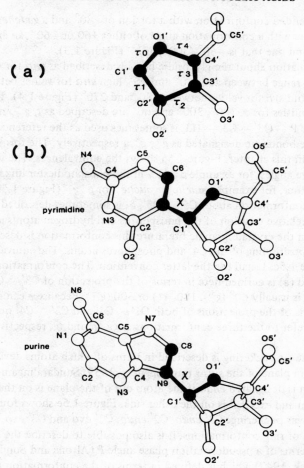

Figure 1.2 Conformational nomenclature used to describe (a) the ribose sugar, (b) the glycosidic bond. The nomenclature is shown for pyrimidine and purine nucleosides

Table 1.1 The various conformational conventions

Bond	1	2	3	4	5	6
$C4'—C3'—O3'—P$	α	ϕ'	ω	ω''	$\theta3$	ϕ'
$C3'—O3'—P—O5'$	β	ω'	ϕ	ψ'	ϕ	ω'
$O3'—P—O5'—C5'$	γ	ω	ψ	ψ''	ψ	ω
$P—O5'—C5'—C4'$	δ	ϕ	θ	ϕ	$\theta1$	ϕ
$O5'—C5'—C4'—C3'$	ϵ	ψ	ζ	ϕ''	$\theta2$	ψ
$C5'—C4'—C3'—O3'$	ζ	ψ'	σ	ω'	σ	ρ'

(1) Seeman *et al.* (1976).
(2) Sundaralingam (1969).
(3) Arnott (1970).
(4) Olson and Flory (1972).
(5) Lakshminarayan and Sasisekaran (1970).
(6) Sussman *et al.* (1972).

denotes an extended conformation with a torsion of $180°$ and a *gauche* (g) conformation is one with a conformation angle of either +60 or $-60°$; an angle of $60°$ is denoted g^+ and one that is $-60°$ ($300°$) as g^- (Figure 1.3).

The conformation about the glycosidic bond is described as *anti* for values of that angle that range between about $0°$ and $75°$, high *anti* for values between about $75°$ and $110°$ and *syn* for values between $180°$ and $270°$ (Figure 1.4). The three rotomer possibilities for α, $180°$, $300°$ and $60°$ are described as t, g^- and g^+. In the NMR literature P—O3'—C3'—H3' is sometimes used as the reference in which case these same bonds are designated as g^-, g^+, t respectively. The former conventions are used in this chapter. Figure 1.5a shows the equivalences. The values of two consecutive angles, for example β and γ in the phosphodiester linkage can be described together, for example *gauche⁻, gauche⁻* or g^-g^- (Figure 1.5b). As in the case of α, the conformation about C5'—O5', δ, is sometimes described with respect to the relative position of the phosphorus and hydrogen atoms as $g'g'$, gt', tg'. However, in the crystallographic literature this conformation is described with respect to the positioning of the C4' and phosphorus atoms. The equivalences are shown in figure 1.5c. I shall use the latter convention. The conformation about the C4'—C5' bond (ϵ) is defined here in terms of the projection of C3'—C4' on C5'—O5' and is usually $60°$ (g^+), $180°$ (t) or 300 (g^-). Because ϵ can also be defined in terms of the projections of both O1'—C4' and C3'—C4' on C5'—O5', some authors refer to the three conformations as gg, gt and tg, respectively (Figure 1.5d).

The ribose sugar puckering is described in terms of which atoms deviate from the best 3 or 4 atom plane of the sugar ring (Spencer, 1959; Sundaralingam, 1965). Hence a conformation is designated *endo* if the atom out of the plane is on the same side as the C5' atom and *exo* if it is on the other side. Figure 1.5e shows four typical examples of sugar puckering: C3' *endo*, C2' *endo*, C2' *exo* and C3' *exo*. Since there is a continuum of ring conformations, it is also possible to describe the ring conformation in terms of a pseudorotation phase angle P (Altona and Sundaralingam, 1972; Kilpatrick, 1947) which is defined in terms of the conformation angles of the ribose ring:

$$\tan P = \frac{(\tau_4 + \tau_1) - (\tau_3 + \tau_0)}{2\tau_2 (\sin 36° + \sin 72°)} \quad \begin{array}{l} \text{(note: when } \tau_2 \text{ is negative} \\ \text{add } 180° \text{ to } P) \end{array}$$

The maximum degree of pucker τ_m is defined by:

$$\tau_2 = \tau_m \cos P$$

Flattened rings have low values of τ_m, for example $20°$, and very puckered rings have large values such as $42°$. In this nomenclature C3' *endo*, C2' *exo* conformations are type N (the northern half of the pseudorotation circle) and centred around $P = 0°$; C2' *endo*, C3' *exo* conformations are type S (the southern half of the pseudorotation circle) centred around $P = 180°$. Symmetrical twist (T) conformations occur at even multiples of $P = 18°$. Figure 1.6 shows the pseudorotation pathway.

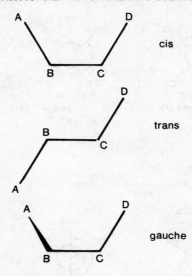

Figure 1.3   Examples of *trans*, *cis* and *gauche* conformations

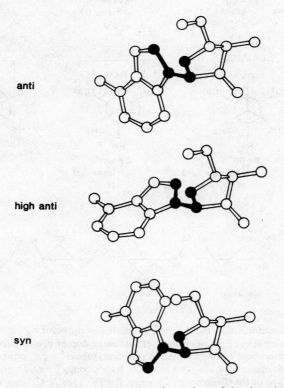

Figure 1.4   *Syn*, *anti* and high *anti* conformations of the glycosidic bond

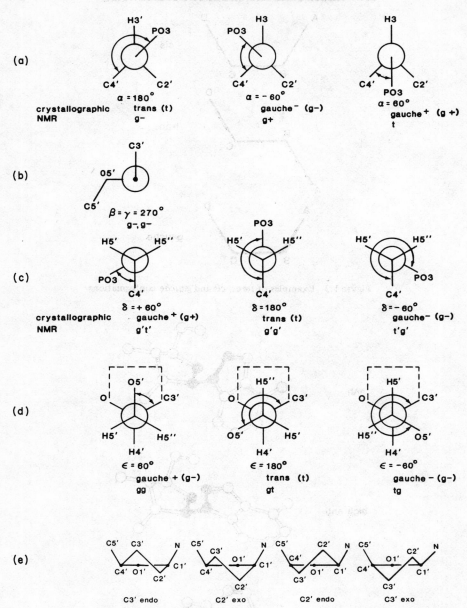

Figure 1.5(a) The conformation about α: two alternate descriptions $t = g^- \; g^- = g^+ \; g^+ = t$. (b) The conformation about β, γ: an example of the description of two consecutive bonds. (c) The conformation about δ: two alternate descriptions $gauche^+ = g't'$, $trans = g'g'$, $gauche^- = t'g'$. (d) The conformation about ϵ: two alternate descriptions $g^+ = gg$, $t = gt$, $g^- = tg$. (e) The conformation of the ribose sugar: C3' $endo$, $P = 18°$, type N; C2' exo, $P = 342°$, type N; C2' $endo$, $P = 162°$, type S; C3' exo, $P = 198°$, type S

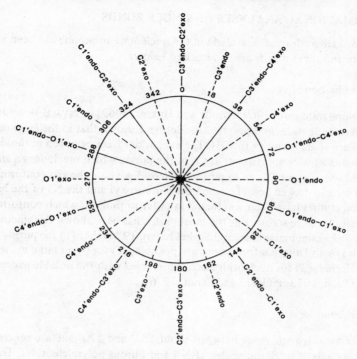

Figure 1.6 The pseudorotation pathway

METHODS

There have been several approaches to conformational analysis of nucleic acids. The two most popular experimental methods are solution studies utilising nuclear magnetic resonance (Wood *et al.*, 1973; Sarma *et al.*, 1974; Altona, 1975; Lee *et al.*, 1976), discussed in chapters 3 and 6, and x-ray diffraction studies of both fibres and single crystals (Arnott, 1970; Kallenbach and Berman, 1977), discussed in chapters 4 and 5. In this latter approach, conformational angles from hundreds of crystals (see for example Sundaralingam, 1969; Arnott and Hukins, 1969), or in the case of tRNA (Holbrook *et al.*, 1978; Jack, Ladner and Klug, 1976) from the 76 linked nucleotides in the same molecule, are analysed for structural correlations between different angles. The theoretical studies have centred on semi-empirical potential energy calculations (Lakshminarayanan and Sasisekharan, 1969*a, b*; Olson and Flory, 1972; Broyde *et al.*, 1975) and quantum mechanical calculations (Pullman and Saran, 1976); these have been reviewed and compared to the experimental results by Olson (1975). Some important features and trends among the conformational angles have emerged from these experimental and theoretical studies although the goal of elucidating firm conformational principles so far has not been met.

CONFORMATIONAL ANALYSES OF SINGLE BONDS

Each individual conformational angle in the nucleotide backbone has been analysed extensively and these results are summarised below.

The glycosidic bond (χ)

Surveys (Sundaralingam, 1969; Arnott and Hukins, 1969) of crystal structures of nucleosides, nucleotides and polynucleotides have shown that in the large majority of cases the χ angle is *anti* (= 0–90°). The exceptions have occurred in those cases where there is a bulky substituent at C8 of the purine which precludes an *anti* conformation, and in some C2′ *endo* purine nucleosides where the *syn* conformation is stabilised by a hydrogen bond between the 5′ hydroxyl and the N3 of the base ring. Theoretical studies have been conducted in order to find out which conformations are intrinsically more stable. For the most part it has been shown (Haschemeyer and Rich, 1967; Lakshminarayanan and Sasisekharan, 1970) that (1) for purine nucleosides both *syn* and *anti* conformations are possible, (2) for pyrimidine nucleosides *anti* is preferred, (3) for nucleotides *anti* is preferred with the notable exception of 5′ GMP (Olson, 1973; Pullman and Saran, 1976).

C3′—O3′ bond: α

The values for this angle range between about 180° and 270° and are centred at 220° for crystals of small molecules, tRNA and fibrous polynucleotides. Theoretical studies (see summary by Olson, 1975) concur with one another and with the experimental results.

The phosphodiester linkage O3′——P, P——O5′: β, γ

The relatively low potential energy barriers to rotations of these two angles coupled with the scarcity of crystallographic data make derivation of conformational principles for the phosphodiester linkage difficult. Whereas dinucleotide and trinucleotide crystal structures (Kallenbach and Berman, 1977) have displayed only three conformations, tg^-, g^-g^- and g^+g^+, and fibrous polynucleotides are almost exclusively g^-g^-, tRNA displays all pairs (Jack, Ladner and Klug, 1976; Holbrook *et al*., 1978) of staggered values for these angles except tt, though most of the residues are g^-g^-. As shown by Olson (1975) there is relatively poor agreement among the various energy estimates. For example, only the PCILO (Pullman, Perahia and Saran, 1972) and the *ab initio* calculations (Newton, 1973) show the relative unimportance of the tt conformation whereas the same empirical and quantum mechnical calculations attach more significance to this conformation than to the experimentally observed tg^- one. Because it has been supposed that most of the flexibility of the polynucleotide chain resides in these angles (Kim *et al*., 1973; Yathindra and Sundaralingam, 1974), a firmer understanding of their rotational constraints is certainly needed.

The O5′—C5′ bond: δ

In most diffraction studies this angle is *trans* (180°) within a fairly broad range. A few residues of the orthorhombic form of tRNA (Holbrook *et al.*, 1978) display the *gauche* conformation. This significant preference for *trans* is also predicted by most of the theoretical calculations.

The C5′—C4′ bond: ε

All three of the potential energy minima, g^-, t, g^+ (Pullman and Saran, 1976; Olson, 1975), have been observed experimentally. Crystals of nucleosides are predominantly g^+ with some examples of t and g^- conformations. There are very few nucleotides with conformations other than g^+; in the structures of 6-azauridine-5′ phosphoric acid (Saenger and Suck, 1973), 5′dGMP (Young *et al.*, 1974), and in ApA complexed with proflavine (Neidle *et al.*, 1978) ε is t. In tRNA, as pointed out by Jack, Klug and Ladner (1976), there are nucleotides with the t as well as the rare g^- conformation. Interestingly, in the unrefined model of DNA (Crick and Watson, 1954), this bond was *trans* as it is in one model of 'kinked' DNA (Crick and Klug, 1975). PCILO (Pullman *et al.*, 1972) and classical energy (Olson and Flory, 1972) calculations predict g^+ and t to be the most likely conformations, as the above crystallographic frequencies appear to support.

The ribose sugar ring: ζ

X-ray studies show that the ribose ring almost always adopts the C3′ *endo* ($\zeta \simeq 75°$) or the C2′ *endo* ($\zeta \simeq 142°$) conformation. In polyribonucleotides the sugars are C3′ *endo* whereas in polydeoxynucleotides they can be either C3′ *endo* or C2′ *endo* (Arnott *et al.*, 1975). (In pseudorotation terminology *P* falls between 0° and 18° for C3′ *endo* and between 144° and 180° for C2′ *endo* (Sundaralingam, 1974).) It has been supposed that the energy barrier of interconversion between these two forms is about 2 or 3 Kcal mole^{-1} (Sundaralingam, 1974; Sasisekharan, 1973), but recent potential energy calculations (Levitt and Warshel, 1978) suggest that this barrier is only about 0·5 Kcal mole^{-1}. These authors invoke packing forces to explain the clustering of the crystallographic results and warn against the use of fixed ribose rings in model building studies. More recent studies reconfirm the 2 to 3 Kcal mole^{-1} barrier (Olson and Sussman, personal communication).

CONFORMATIONAL CORRELATIONS BETWEEN BONDS

Not unexpectedly, the conformation angles in a nucleotide can sometimes be correlated with one another. Some of these 49 possible correlations are now described.

Glycosyl – sugar pucker: χ, ζ

In crystals of nucleosides and nucleotides the values of χ for C3′ *endo* sugars are lower than for C2′ *endo* sugars which usually have high *anti* conformations. This is further corroborated by the values of the conformation angles in fibrous poly-nucleotides (Arnott *et al.*, 1975) and in tRNA (Holbrook *et al.*, 1978; Jack, Ladner and Klug, 1976). Notable exceptions to this occur in the structures of some of the drug–dinucleoside complexes where combinations of 'high' χ and C3' *endo* conformations are observed (Seeman *et al.*, 1975; Neidle *et al.*, 1978; Berman *et al.*, 1979).

Glycosyl – C5'—C4': χ, ϵ

In crystals of nucleotides with either C3' *endo* or C2' *endo* sugars, the preferred conformation is *anti* for the χ angle coupled with g^+ for ϵ (Sundaralingam, 1973; Yathindra and Sundaralingam, 1973). In purine nucleosides the *syn- g^+* conformation has some importance since it is possible to form a hydrogen bond between the 5' hydroxyl and the N3 of the base. There is evidence from NMR studies that for purine nucleotides a *syn* conformation is correlated with a g^- or *t* conformation for ϵ (Sarma *et al.*, 1974). Theoretical calculations (Olson, 1973) have shown that in poly rA with C3' *endo* sugars, the possible χ, ϵ combinations are *anti-t*, *anti-g^+* and *syn-t* and for C2' *endo* sugars *anti-t*, *anti-g^+*, *syn-t* and *syn-g^+* are possible with *anti-g^+* preferred. In these calculations g^- conformations are shown to be forbidden for polynucleotides.

Sugar pucker – C5'—C4': ζ, ϵ

In nucleotides and nucleosides with C3' *endo* sugars, ϵ is either g^+ or *t*. For steric reasons, g^- conformations for ϵ are only possible for C2' *endo* sugar puckers as has been demonstrated in tRNA (Jack, Klug and Ladner, 1976).

Sugar pucker – C3'—O3': ζ, α

For most nucleotides with C3' *endo* pucker, γ is about 210° and in those nucleo-tides with C2' *endo* sugar puckers the α values are 210° or 270°.

C5'—C4', O5'—P: ϵ, γ

In tRNA it has been shown that as β increases in value, ϵ decreases proportionally (Holbrook *et al.*, 1978). This correlation has been ascribed to base stacking stabilisa-tion, that is, when one angle changes, stacking is maintained by changing the coupled angle. A similar effect is noted by Broyde *et al.* (1975) in their studies of dGpdC where they point out that ϵ, γ = 173°, 158° (the original Watson-Crick con-formation) is equivalent energetically to ϵ, γ = 50°, 297°.

O5'—C5', O3'—P: δ, β

Holbrook *et al.* (1978) show another example of alternate bond couplings for the O5'—C5' and O3'—P bonds in tRNA. As ϵ increases, so does α.

C5′—C4′, C3′—O3′: ϵ, α

The strong correlation in tRNA is displayed by the ϵ and α bonds; as ϵ increases, so does α.

C5′—C4′, O5′—C5′: ϵ, δ

Theoretical studies by Olson and Flory (1972) have demonstrated the interdependence of ϵ and δ. NMR studies by Wood *et al.* (1973) show that when ϵ is g^+ then only the t conformation for δ is possible, and that when ϵ is t or g^- the other rotomers of δ are possible but not highly likely. In most x-ray structures of small and large molecules, δ is *trans* no matter what conformation ϵ assumes. However, in the drug–dinucleoside complexes, in which ϵ is invariably g^+, the value of δ can only *very* loosely be described as *trans* (Berman *et al.*, 1978).

THE RIGID NUCLEOTIDE CONCEPT

The first attempts to reduce the apparent complexity of nucleic acid conformation came in 1969 when Arnott and Hukins (1969) and Sundaralingam (1969) surveyed the known crystal and fibre structures of nucleic acids and their constituents and discovered that the conformational angles were confined to rather narrow ranges of

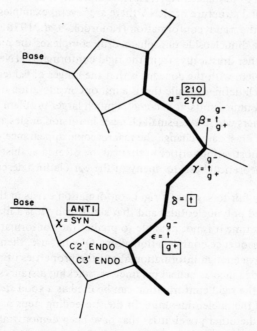

Figure 1.7   The nucleotide backbone showing the preferred conformations

values. In nucleosides χ is either *syn* or *anti*, the sugar pucker is either C2′ *endo* or C3′ *endo*, α is *trans*, δ is *trans* and ε is *gauche⁺*, *trans* or *gauche⁻* (Figure 1.7). In polynucleotides the preferred conformation for χ is *anti* and for ε, *gauche⁺*. In the first crystal structure determination of a dinucleoside phosphate (Sussman *et al.*, 1972; Rubin *et al.*, 1972) the nucleotide units showed the same conformational features as those in fibrous helical polynucleotides. As a result, the concept of the 'rigid' nucleotide arose which stated that nucleotides are conformationally more restricted than nucleosides and that the conformational flexibility of polynucleotides resides in the phosphodiester linkage, β and γ, as was first suggested by Arnott and Hukins (1969). The value of such a concept is obvious since, if true, it would greatly simplify model building of single-stranded nucleic acids and in fitting low resolution Fourier maps of these molecules. As a direct consequence of the development of this concept a great deal of effort was expended in trying to define the conformations of the phosphodiester linkage (Yathindra and Sundaralingam, 1974; Kim *et al.*, 1973; Olson, 1975), and to build models of polynucleotides by varying the torsion angles of this linkage alone.

Several criticisms have been raised about the validity of the concept. It is definitely not clear that nucleotides are more 'rigid' than nucleosides. For example, Berthod and Pullman (1973) have demonstrated with theoretical calculations that in the nucleotide 5′ GMP the *syn* conformation is more stable than the *anti* conformation. In solution studies, Evans and Sarma (1976) demonstrated that the nucleotidyl units of destacked ApA, ApApA and polyA are as flexible as the nucleosides. In the crystal structure of tRNA there are several examples of nucleotides with other than the 'rigid' conformation (Holbrook *et al.*, 1978; Hingerty *et al.*, 1978), and in the dinucleoside phosphate–drug complexes the nucleotides deviate, in some cases rather drastically, from the rigid conformation (Neidle *et al.*, 1978).

Another problem with the concept is that the ranges of values for the torsion angles are not well defined. While this is a minor consideration on building models of monomer and dimer units, it becomes a much larger problem when building models of polymers where even small changes in torsion angles are propagated in such a way as to drastically change the macroscopic appearance of the molecule.

Finally, the most serious criticism that can be offered against the concept is the lack of a firm theoretical basis for many of the correlations described in previous sections.

It is probably fair to say that a rigid conformation exists in the nucleotide units of helical stacked polynucleotides and that statistically it is a likely conformation. However, at this time it is not possible to predict the conformation of any single molecule given sequence conformation alone. Of what use, then, is the concept? Assuming we have enough information about the properties of a particular oligo or polynucleotide (such as helical parameters, stacking distances or proton–proton distances) then the rigid conformation can be used as a good *starting* approximation for the shape of the nucleotide units. In the succeeding steps of the model building, however, all of the other possibilities that have been demonstrated experimentally and predicted theoretically must not be excluded.

HELICES, KINKS AND LOOPS

The concentration of efforts in recent years on the structure of chromatin and on the primary structures of various DNA and RNAs has produced some interesting and challenging conformational problems.

For example, how does DNA wind around a histone core? It has long been known that double stranded DNA is helical (Watson and Crick, 1953) and detailed studies (Arnott, 1975) have provided accurate conformational angles for a wide variety of polydeoxynucleotide helices. These helices are characterised by each nucleotide unit having the same conformation, which may be qualitatively described as $\alpha = t$, $\beta = g^-$, $\gamma = g^-$, $\delta = t$, $\epsilon = g^+$ and $\zeta = t$ or g^+ (C2' *endo* or C3' *endo*). The structure of this class of polymer molecules is thus necessarily a long straight tube. There are several ways to deform this tube or rod so that it can bend around a protein. One way suggested by Crick and Klug (1975) is to introduce a *kink*, say every ten base pairs, by rotating 120° about the C5'—C4' and thus changing ϵ from a g^+ to t conformation. However, it is also possible to smoothly deform DNA, by rather more gradual subtle changes in the sugar phosphate backbone as has been demonstrated by both Levitt (1978) and Sussman and Trifonov (1978).

Of concern in the understanding of single-stranded nucleic acids is the elucidation of principles which guide the formation of loops and bends. The crystal structure of tRNA has provided data that should be useful towards formulating those principles. For example, Kim and Sussman (1976) have noted that the TψC and anti *codon* loops bend in similar ways. They define π turns (also called U turns by Quigley and Rich, 1976) as turns due primarily to rotations about β and γ, the phosphodiester linkage. The values for β of π_1, π_2 and π_3 turns are 85°, 170° and 285°, respectively. The sharpness of the turn depends on the deviation of γ from 290°.

Obviously, there is a long road ahead in the understanding of nucleic acid conformations. As more data become available and more sophisticated methods are developed to interpret these data the goal of elucidating a conformational code for polynucleotides should be obtainable.

ACKNOWLEDGMENTS

We acknowledge the support of grants from NIH: GM-21589, CA-22780, CA-06927, RR-05539 and an appropriation from the Commonwealth of Pennsylvania.

REFERENCES

Altona, C. (1975). In *Structure and Conformation of Nucleic Acids and Protein–Nucleic Acid Interactions* (ed. M. Sundaralingam and S. T. Rao), University Park Press, Baltimore, p. 613.

Altona, C. and Sundaralingam, M. (1972). *J. Am. chem. Soc.,* **94**, 8205.

Arnott, S. (1970). *Prog. Biophys. Mol. Biol.,* **21**, 265.

Arnott, S., Chandrasekaran, R. and Selsing, E. (1975). In *Structure and Conformation of Nucleic Acids and Protein–Nucleic Acid Interactions* (ed. M. Sundaralingam, and S. T. Rao), University Park Press, Baltimore, p. 577.

Arnott, S. and Hukins, D. W. L. (1969). *Nature, Lond.,* **224**, 886.

Berman, H. M., Neidle, S. and Stodola, R. K. (1978). *Proc. natn. Acad. Sci. U.S.A.,* **75**, 828.
Berman, H. M., Stallings, W., Carrell, H. L., Glusker, J. P., Neidle, S., Taylor, G. and Achari, A. (1979). *Biopolymers,* **18**, 2405.
Berthod, H. and Pullman, B. (1973). *FEBS Lett.,* **30**, 231.
Broyde, S. B., Wartell, R. M., Stellman, S. D., Hingerty, B. and Langridge, R. (1975). *Biopolymers,* **14**, 1597.
Crick, F. H. C. and Klug, A. (1975). *Nature, Lond.,* **255**, 530.
Crick, F. H. C. and Watson, J. D. (1954). *Proc. R. Soc. A,* **223**, 80.
Evans, F. E. and Sarma, R. H. (1976). *Nature, Lond.,* **263**, 567.
Haschemeyer, A. E. V. and Rich, A. (1967). *J. molec. Biol.,* **27**, 369.
Hingerty, B., Brown, R. S. and Jack, A. (1978). *J. molec. Biol.,* **124**, 523.
Holbrook, S. R., Sussman, J. L., Warrant, R. W. and Kim, S. H. (1978). *J. molec. Biol.,* **123**, 631.
Jack, A., Klug, A. and Ladner, J. E. (1976). *Nature, Lond.,* **261**, 250.
Jack, A., Ladner, J. E. and Klug, A. (1976). *J. molec. Biol.,* **108**, 619.
Kallenbach, N. and Berman, H. M. (1977). *Q. Rev. Biophys.,* **10**, 138.
Kilpatrick, J. E., Pitzer, K. S. and Spitzer, R. (1947). *J. Am. chem. Soc.,* **69**, 2483.
Kim, S. H., Berman, H. M., Seeman, N. C. and Newton, M. D. (1973). *Acta crystallogr.,* **B29**, 703.
Kim, S. H. and Sussman, J. L. (1976). *Nature, Lond.,* **260**, 654.
Lakshminarayan, A. V. and Sasisekharan, V. (1969a). *Biopolymers,* **8**, 475.
Lakshminarayan, A. V. and Sasisekharan, V. (1969b). *Biopolymers,* **8**, 489.
Lakshminarayan, A. V. and Sasisekharan, V. (1970). *Biochim. biophys. Acta,* **204**, 49.
Lee, C. H., Ezra, F., Kondo, N. S., Sarma, R. H. and Danyluk, S. S. (1976). *Biochemistry,* **15**, 3627.
Levitt, M. (1978). *Proc. natn. Acad. Sci. U.S.A.,* **75**, 640.
Levitt, M. and Warshel, A. (1978). *J. Am. chem. Soc.,* **100**, 2607.
Neidle, S., Taylor, G., Sanderson, M., Shieh, H-S. and Berman, H. M. (1978). *Nucleic Acids Res.,* **5**, 4417.
Newton, M. D. (1973). *J. Am. chem. Soc.,* **95**, 256.
Olson, W. K. (1973). *Biopolymers,* **12**, 1787.
Olson, W. K. (1975). *Biopolymers,* **14**, 1775.
Olson, W. K. and Flory, P. J. (1972). *Biopolymers,* **11**, 1.
Pullman, B., Perahia, D. and Saran, A. (1972). *Biochim. biophys. Acta,* **269**, 1.
Pullman, B. and Saran, A. (1976). *Prog. Nucl. Acid Res. Mol. Biol.,* **18**, 215.
Quigley, G. L. and Rich, A. (1976). *Science,* **194**, 796.
Rubin, J., Brennan, T. and Sundaralingam, M. (1972). *Biochemistry,* **11**, 3112.
Saenger, W. and Suck, D. (1973). *Nature, Lond.,* **242**, 610.
Sarma, R., Lee, C. H., Evans, F. E., Yathindra, N. and Sundaralingam, M. (1974). *J. Am. chem. Soc.,* **96**, 7337.
Sasisekharan, V. (1973). *Conformation of Biological Molecules and Polymers, 5th Jerusalem Symp. Quant. Chem. Biochem.* (ed. E. D. Bergmann and B. Pullman), Academic Press, New York.
Seeman, N. C., Day, R. O. and Rich, A. (1975). *Nature, Lond.,* **253**, 329.
Seeman, N. C., Rosenberg, J. M., Suddath, F. L., Kim, J. J. and Rich, A. (1976). *J. molec. Biol.,* **104**, 109.
Spencer, M. (1959). *Acta crystallogr.,* **12**, 59.
Sundaralingam, M. (1965). *J. Am. chem. Soc.,* **87**, 599.
Sundaralingam, M. (1969). *Biopolymers,* **7**, 821.
Sundaralingam, M. (1973). *Conformation of Biological Molecules and Polymers, 5th Jerusalem Symp. Quant. Chem. Biochem.* (ed. E. D. Bergmann, and B. Pullman), Academic Press, New York, p. 417.

Sundaralingam, M. (1974). *Int. J. Quantum Chem. Quantum Biol. Symp.,* **1**, 81.
Sussman, J. L., Seeman, N. C., Kim, S. H. and Berman, H. M. (1972). *J. molec. Biol.,* **66**, 403.
Sussman, J. L. and Trifenov, E. N. (1978). *Proc. natn. Acad. Sci. U.S.A.,* **75**, 103.
Watson, J. D. and Crick, F. H. C. (1953). *Nature, Lond.,* **171**, 737.
Wood, D. J., Mynott, R. J., Hruska, F. B. and Sarma, R. H. (1973). *FEBS Lett.,* **34**, 323.
Young, D. W., Tollen, P. and Wilson, H. R. (1974). *Nature, Lond.,* **248**, 513.
Yathindra, N. and Sundaralingam, M. (1973). *Biopolymers,* **12**, 297.
Yathindra, N. and Sundaralingam, M. (1974). *Proc. natn. Acad. Sci. U.S.A.,* **71**, 3325.

2
Crystal structures of fragments of DNA and RNA

Helen M. Berman and Huey-Sheng Shieh

INTRODUCTION

Crystallographic analyses of oligonucleotides give the degree of detailed information about nucleic acid structure that is not obtainable from fibrous polynucleotides, which yield average structures, and globular nucleic acids such as tRNA which yield relatively low resolution data. Indeed, not until 1976 was the Watson-Crick geometry of an A-U base pair actually observed at atomic resolution, in the crystal structure of ApU (Seeman *et al*., 1976). The smallest unit of a polynucleotide which contains all the conformational information necessary to describe the relationship among the base, sugar and phosphodiester linkage is the dinucleoside phosphate. Since it was first demonstrated that it is possible to obtain good crystals of these molecules (Seeman *et al*., 1971), considerable effort has been expended in studying oligonucleotides alone and complexed with other molecules. This chapter concentrates on the uncomplexed oligonucleotides and summarises what is now known about their conformations, hydrogen bonding properties, stacking patterns and metal binding sites.

THE CRYSTALLOGRAPHIC EXPERIMENT

As shown in table 2.1, there have been relatively few crystal structure determinations of nucleic acid fragments. The first structure determination of a dinucleoside phosphate occurred in 1971 (Seeman *et al*., 1976). The first tetranucleotide (dpApTpApT) structure analysis occurred in 1978 (Viswamitra *et al*., 1978). The structures of the deoxy hexamer dCpGpCpGpCpG (Wang *et al*., 1979) and the deoxy tetramer dCpGpCpG (Drew *et al*., 1978) are the most recent additions to a slowly growing list. Structural details were not available at the time of writing, and so are not presented in the tables and figures. The reasons for this are that pure homogeneous materials necessary for producing crystals are not easy to obtain, and crystallisation of x-ray grade crystals is difficult. When crystals are obtained and data collected, the determination of the coordinates of each atom in the asymmetric unit of the crystal is not always straightforward for the following reasons.

17

Table 2.1 Crystallographic details of oligonucleotide structures

Structure	Number of nucleotides	Water molecules Full	Partial	Metal ions	Resolution (Å)	Number of reflections All	Observed	R factor (observed data)	Estimated standard deviations Co-ordinates	Bond lengths (Å)	Bond angles (°)	Reference
NaGpC $C_{19}H_{24}N_8O_{12}P$	1	8	2	$1Na^+$	0.9	2630	2276	0.05	0.007	0.01	0.8	Rosenberg et al. (1976)
Ca(GpC)$_2$ $C_{19}H_{24}N_8O_{12}P$	4	32	8	$2Ca^{2+}$	1.1	4237	2918	0.08	0.02	0.03	1.5	Hingerty et al. (1976)
NaApU $C_{19}H_{24}N_8O_{12}P$	2	10	4	$2Na^+$	0.8		4700	0.06	0.005	0.008	0.5	Seeman et al. (1976)
UpA $C_{19}H_{23}N_7O_{12}P$	2	1	0	–	1.0	2680	2283	0.06	0.007	0.015	1.0	Sussman et al. (1972)
$C_{19}H_{24}N_7O_{12}P$					0.9	4024	3179	0.04	0.005	0.008	0.5	Rubin et al. (1972)
NadTpT $C_{20}H_{25}N_4O_{15}P_2$	1	10	5	$2Na^+$	1.2	1373	971	0.11	0.03	0.035	2.5	Camerman et al. (1976)
ApApA $C_{30}H_{37}N_{15}O_{16}P_2$	1	4	4		0.9	3260	3210	0.06	0.007	0.005	0.3	Suck et al. (1976)
dpApTpApT $C_4H_{48}N_{14}O_{25}P_4$	1	31/52 sites			1.0	4258	2717	0.15	0.03	0.04	3.0	Viswamitra et al. (1978)

(1) There are many atoms in the crystal.

(2) The atoms are not 'randomly' distributed since there are many stacked base planes and, therefore, 'direct' statistical methods of phase determination are usually not successful.

(3) The crystals sometimes have pseudosymmetry.

Some of the structures have been determined by taking advantage of the fact that the phosphorus atom is a stronger x-ray scatterer than the rest of the atoms and thus can be located more easily and used to phase the other atoms in the crystal. Refinement of these structures can be difficult because some of the crystals are highly hydrated so that it is not always clear whether residual Fourier peaks represent disordered water molecules or errors in the model.

DESCRIPTION OF THE STRUCTURES

Geometry of the nucleotides

The dinucleoside phosphate portions of most oligomeric crystal structures studied thus far fall into three conformational classes (table 2.2). The self-complementary

Table 2.2 Conformation angles ($^\circ$) in oligomeric structures

	α	β	γ	δ	ϵ	$\zeta(5')$	$\zeta(3')$	$\chi(5')$	$\chi(3')$	Reference
GpC	211	292	285	184	50	89	77	13	32	Rosenberg *et al.* (1976)
GpC	222	293	291	181	47	76	79	8	33	Hingerty *et al.* (1976)
	217	291	292	172	57	73	80	4	22	
	224	290	286	167	63	80	74	1	25	
	217	288	282	181	52	88	87	7	28	
ApU	213	293	288	177	57	84	74	7	28	Seeman *et al.* (1976)
	221	284	295	168	58	78	77	4	29	
dA1pT2	212	294	292	176	68	90	134	4	67	Viswamitra *et al.* (1978)
dA3pT4	212	285	300	170	64	83	139	−11	74	
A1pA2	223	283	297	161	56	82	81	7	21	Suck *et al.* (1976)
Average	217(5)	289(4)	290(6)	174(7)	57(7)					
UpA2	224	164	271	192	54	77	93	19	44	Sussman *et al.* (1972)
dT2pA3	203	168	286	187	50	134	80	67	−11	Viswamitra *et al.* (1978)
dTpT	255	164	294	184	42	149	152	32	35	Camerman *et al.* (1976)
UpA1	205	81	82	203	55	86	85	12	37	Sussman *et al.* (1972)
A2pA3	207	76	92	196	56	81	79	21	21	Suck *et al.* (1976)
RNA 11	213	281	300	175	49	83	83	14	14	Arnott *et al.* (1976)
DNA 10	159	261	321	209	31	157	157	85	85	

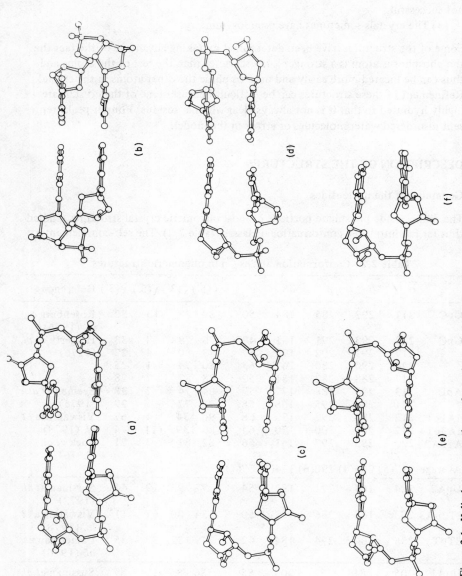

Figure 2.1 Helical structures, the conformations of which resemble those of RNA and DNA: (a) RNA 11, (b) DNA 10, (c) ApU, (d) dApT in dpApTpApT, (e) GpC, (f) A1pA2 in ApApA

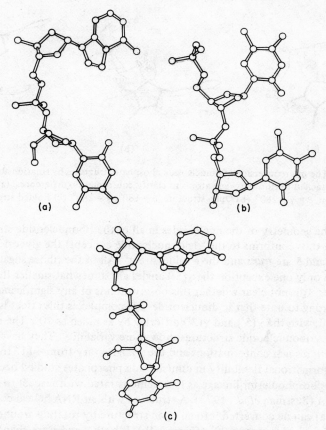

Figure 2.2   Open structures: (a) dTpA and dpApTpApT
(b) dpTpT, (c) UpA2 in UpA

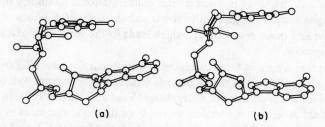

Figure 2.3   Folded structures: (a) A2pA3 in ApApA, (b) UpA1 in UpA

duplexes exemplified by GpC and ApU have been dubbed 'mini helices' because of
their resemblance to fragments of double helical DNA and RNA (figure 2.1). UpA2
and dpTpT form extended structures as shown in figure 2.2. The third class of
structures (figure 2.3), exemplified by UpA1, form folded structures with the bases

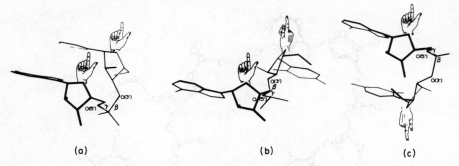

(a)   (b)   (c)

Figure 2.4   The interconversion of dinucleoside phosphate structures by rotation about β and γ. Hands are attached to ribose oxygen atoms to clarify conformation differences. (a) RNA helical conformation, $\beta = \gamma = 280°$. (b) Open structure, $\beta = 180°$, $\gamma = 280°$. (c) Folded structure, $\beta = \gamma = 80°$

stacked. The geometry of the nucleosides in all of the ribonucleoside structures is similar and thus conforms to the 'rigid' nucleotide concept: the glycosidic angles χ are *anti*, α and δ are *trans* and the ϵ values are g^+. All of the ribose sugars are C3' *endo*. With only one exception the $\chi(5')$ angles are somewhat smaller than the $\chi(3')$ angles. It is not clear whether this observation is of any significance although it is interesting to note that in dinucleotide–dye complexes this effect is greatly exaggerated, with the $\chi(5')$ and $\chi(3')$ differing by as much as 70°. The nucleosides in the deoxyribonucleotide structures show more variability. They have one or more C2' *endo* sugar conformations and the χ angles vary from −11° to 74°. The major conformational flexibility in dinucleoside phosphates studied occurs within the phosphodiester linkage as first demonstrated with respect to the structure of UpA (Sussman *et al.*, 1972). A structure with an RNA helical conformation (figure 2.4a) can be converted into an open structure by rotating around the P—O3' (β) bond by about 120° (figure 2.4b). If both β and γ are changed by 180° from the values in the RNA type structure, a folded structure is obtained (figure 2.4c). Interestingly, in the larger hexameric structure, dCpGpCpGpCpG (Wang *et al.*, 1979), there is much greater conformational flexibility in that the torsion angles around the glycosidic bonds and the C4'—C5' bonds of the guanosine are *syn* and *trans*, respectively, which leads to the formation of a left-handed Z-helix fragment.

The average weighted bond distances and angles of the bases (figure 2.5) and the ribose sugars (table 2.3) in the oligomeric structures correspond well to the values reported for base and nucleoside structures (Voet and Rich, 1970). In this compilation there are no apparent correlations between the bond distances or angles and the conformations of the sugars. However, the lack of many C2' *endo* structures makes it difficult to make a firm statement on this point.

Geometry of the base pairs

While G-C base pairs with the Watson-Crick geometry had been observed in the crystal structures of the intermolecular complexes of the component bases alone (Voet and Rich, 1970), it wasn't until the crystal structure of the self-complementary duplex ApU (Seeman *et al.*, 1976) was determined that the Watson-Crick

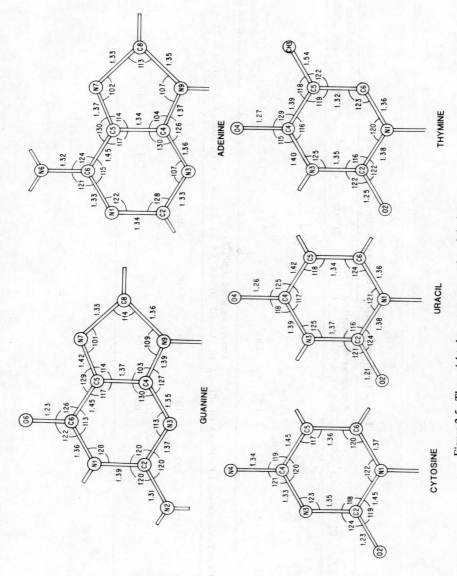

Figure 2.5   The weighted average geometries of the bases

Table 2.3 Ribose geometry in oligonucleotides

	C1'–C2'	C2'–C3'	C3'–C4'	C4'–C5'	C4'–O1'	C1'–O1'	C2'–O2'	C3'–O3'	C1'–N	Number of contributors
					Average weighted distances (Å)					
C2'endo ribose sugars	1.49	1.52	1.53	1.57	1.40	1.47	–	1.48	1.48	4
C3'endo ribose sugars	1.52	1.53	1.52	1.51	1.45	1.41	1.43	1.42	1.49	23

	C1'–C2'–C3'	C3'–C4'–C5'	C4'–O1'–C1'	O1'–C1'–C2'	C2'–C3'–C4'	O1'–C4'–C3'	O1'–C4'–C5'	C2'–C3'–O3'	C4'–C3'–O3'	C1'–C2'–O2'	C2'–O3'–O2'	C3'–C2'–O2'
					Average weighted angles (°)							
C2'endo ribose sugars	105	111	107	107	109	109	102	110	106	–		
C3'endo ribose sugars	101	117	110	108	104	110	102	114	111	107	112	

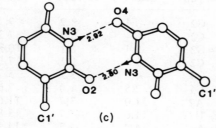

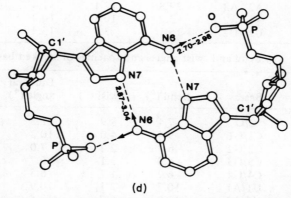

Figure 2.6   Base-pairing geometry: (a) G-C in NaGpC and Ca(GpC)$_2$, (b) A-U in NaApU, (c) U-U in UpA, (d) A-A in UpA, and ApApA

A-U geometry was seen. The crystal structure of dpApTpApT (Viswamitra *et al.*, 1978) shows Watson-Crick A-T pairing. The earlier determination of the crystal structure of UpA (Sussman *et al.*, 1972) did not show this geometry because the N1 atoms of the adenine bases were protonated. Instead, cyclic A-A pairing of the type proposed for acid poly (A^+) (Rich *et al.*, 1961) was observed in this structure and later in the crystal structure of ApApA (Suck *et al.*, 1976). In this kind of bonding the N6 atoms of each adenine base are hydrogen bonded to N7 of another base as well as to a phosphate oxygen. As shown in figure 2.6 the range of hydrogen bonding distances is quite broad with the exception of the N3—N1 bonds whose ranges are narrow. It is not clear whether the shorter distances of the N3—N1 bonds in A-U and A-T pairs relative to G-C distances are of any significance.

The bases in the pairs are not coplanar as shown in figure 2.1 and in table 2.4a. The usual way to describe the distortion is to calculate the dihedral angle between bases in the pair. However, this description is incomplete. The dihedral angle can be decomposed into two components: one is the rotation along the axis parallel to the hydrogen bond, called the *twist* angle and the other one is the rotation along the

Table 2.4
(a) Bending and twisting of base pairs

Structure	Base pair	Bend (°)	Twist (°)	Dihedral angle (°)	Base turn (°)
NaGpC	C1—G2	−6.4	3.2	7.2	30
	G1—C2	6.4	3.2	7.2	
CaGpC	C1—G2	−8.4	6.1	10.3	28
	G1—C2	7.4	−1.2	7.5	
	C3—G4	−4.7	3.2	5.7	29
	G3—C4	7.5	1.0	7.6	
NaApU	U1—A2	−11.0	−3.7	11.6	27
	A1—U2	11.6	−7.7	13.8	
dpApTpApT	T2—A3	−14.2	2.7	14.4	31
	A1—T4	13.4	0.9	13.4	
ApApA	A2—A3	−29.8	12.1	31.4	
UpA	A1—A2	2.2	−10.1	10.3	
	U1—U2	−16.1	−3.3	16.4	

(b) Bend and twist angles and spacings of stacked bases

Structure	Bases	Bend (°)	Twist (°)	Dihedral angle (°)	Spacing (°)
NaGpC	C/G	−7.4	8.9	11.5	3.69
CaGpC	C1/G1	−9.4	4.0	10.2	3.56
	C2/G2	−6.7	−2.2	7.0	3.65
	C3/G3	−6.0	3.1	6.7	3.59
	C4/G4	−6.2	1.7	6.5	3.56
NaApU	U1/A1	−10.7	−2.1	10.9	3.60
	U2/A2	−11.9	−6.7	13.5	3.72
dpApTpApT	T2/A1	−16.2	3.4	16.5	3.51
	T4/A3	−12.8	2.5	13.1	3.53

axis perpendicular to the previous one and called the *bend* angle. Both axes are parallel to the least squares plane of the reference ring (figure 2.7) (see appendix 2.1). These calculations show that in the duplex structures the base pairs are more bent than twisted. As pointed out by Levitt (1978), these distortions can increase the flexibility of polynucleotide structures.

Base stacking

The bases in each strand are not precisely parallel to one another and the dihedral angles between two stacked bases range between 6.5° and 16.5°. Table 2.4b presents

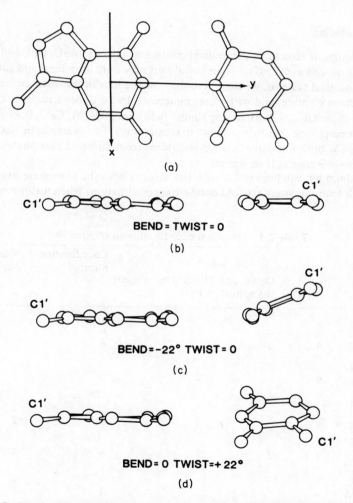

Figure 2.7 Base bending and twisting. (a) A projection view of the chosen coordinate system; Z-axis in this figure goes out of the paper (origin of this coordinate system is unimportant). (b) A coplanar base pair with D_b = 0 and D_t = 0. (c) A bent base pair with D_b = −22° and D_t = 0. (d) A twisted base pair with D_b = 0° and D_t = 22°

the bend and twist as well as dihedral angles for stacked bases. Again, it shows that the bases are more bent than twisted with respect to one another. Furthermore, bend angles are all negative and twist angles have the tendency to be positive (except in ApU which is under the influence of Na coordination, as explained in the following section). The former implies that the head part (see appendix 2.1) of the bottom target base within a strand tilts up; the latter means that the major groove region tilts up. The combined effects bring the polar atoms of the stacked bases close to one another. The stacking distances, ranging from 3.5 to 3.7 Å, are long compared to the values of 3.3 to 3.5 Å in base, nucleoside, or mononucleotide structures (Bugg *et al.*, 1971) and 3.4 Å in RNA and DNA structures (Arnott, Smith and Chandrasekaran, 1976).

Metal ion binding

Several oligomeric structures determined contain metal ions: ApU, GpC and dpTpT (table 2.5), as well as $d(CpG)_3$. Since metal ions play both structural and catalytic roles in biological systems, we shall describe how the ions in these structures interact with the nucleotides and with water molecules. In the dimer structures there are two kinds of metal ions, both having similar radii: Na^+, 0.97 Å, Ca^{2+}, 0.99 Å, and so it is not surprising that they interact in similar ways. For example, in NaGpC and $Ca(GpC)_2$ both ions have two oxygen atoms from different phosphates and four water molecules as their ligands.

The sodium ion can bind either with the bases or with the phosphate oxygen atoms with approximately six-fold octahedral coordination. When water molecules

Table 2.5 Metal ion coordination in oligomers

		Ligands			Coordination number	Number of nucleotides
	H_2O	O_2 (uracil thymine)	Phosphate oxygens Free	Ester		
NaApU						
$Na^+(1)$	2	—	2	2	6	2
$Na^+(2)$	4	2	—	—	6	2
NaGpC						
Na^+	4	—	2	—	6	2
$Ca(GpC)_2$						
$Ca^+(1)$	4	—	2	—	6	2
$Ca^+(2)$	4	—	2	—	6	2
NadpTpT						
$Na^+(1)$	3	2	1	—	6	3
$Na^+(2)^a$						
0.5	6	—	—	—	6	0
0.25	4	—	—	—	4	0
0.25	3	1	1	—	5	2

[a]Disordered

are the only ligands, they tend to be disordered and to adopt tetrahedral coordination as with the second Na^+ in the dpTpT structure.

In structures that contain uracil or thymine, the O2 atoms interact with the metal ion. In the structure of NaApU, in which O2 of the uracil is not involved in Watson-Crick bonding, the Na^+ coordinates with two uracil O2s in the minor groove of the duplex. This type of binding would provide an additional connection between two strands in a double helical structure. As pointed out by Seeman et al. (1976), this coordination depends on the base sequence in the oligonucleotide. In the duplex ApU structure, the distance between the O2 atoms of the stacked uracil groups is ~ 4 Å, whereas in a duplex of UpA the separation would be ~ 8 Å making Na^+ coordination to the bases impossible. In the single-stranded structure of NadpTpT the Na^+ coordinates with the O2 atoms of two symmetry-related thymine groups. In both cases, the sodium ion slips between the planes of two bases and interacts with the bases above it and below it.

The phosphate group is the other binding site for metal ions. NaApU is the only example so far in which the Na^+ binds both to an esterified oxygen (O3') and a nonesterified oxygen (O1) of a phosphate group. In the other cases, the ionic coordination of the Na^+ or Ca^{2+} occurs with the nonesterified oxygen atoms only. Since the coordinations of metal ions in general are not limited to one nucleotide strand, they may indeed serve to stabilise the crystal structure.

The effects of metal ions on nucleotide conformation seem to be minor. In ApU, because the Na^+ binds to two O2s of stacked uracil groups, it appears to affect the base twist angle. When Na^+ coordinates to the O3'A and O1 of the ApU phosphate group, the bond angle O3'A—P—O1 seems to decrease. Different ions may perturb the structure indirectly, but if the two structures NaGpC and $Ca(GpC)_2$ are compared, the conformational differences appear to be small.

MODEL BUILDING

The structures of the oligomers provide fertile ground for model building. For example, it is possible to extract helical parameters from fragments of polynucleic acids by constructing 'helix probes' from equivalent atoms in sequential residues (Rosenberg et al., 1976b) or directly from the six backbone torsion angles (Hingerty, 1979). As shown in table 2.6, the helical parameters calculated for the self-complementary structures by these two methods are systematically different. Furthermore, Hingerty has shown that small changes in the backbone torsion angles and bond angles can cause rather large variations in the helical parameters. These types of change when propagated in a polymer can cause large changes in its overall shape and hence its recognition properties. For example, a single nine-fold helical poly A structure has been constructed from the $A1pA2^+$ fragment of ApApA (Suck et al., 1976), but an eight-fold helix built by Hingerty allows for the same type of hydrogen bonding found in double stranded poly A^+ (Rich et al., 1961). UpA2 (Sussman et al., 1972) and dTpT (Camerman et al., 1976) can be formed into single helices with the bases on the outside and the phosphates on the inside. The very different helical parameters of these two polymer models can be

Table 2.6 Helical parameters for some oligomers

	Translation (Å)	Rotation (°)	Residues per turn	Reference
NaGpC	2.6, 2.4	35, 44	10.4, 8.2	Rosenberg *et al.* (1976)
Ca(GpC)₂				
(average)	2.5, 2.6	31, 46	11.5, 7.8	Hingerty *et al.* (1976)
NaApU	2.4, 3.0	30, 40	11.9, 9.0	Seeman *et al.* (1976)
UpA2	3.1, 3.3	91, 88	4.0, 4.1	Sussman *et al.* (1972)
A1p⁺A2	2.8, 2.7	40, 44	9.0, 8.1	Suck *et al.* (1976)
RNA 11	2.8	33	11.0	Arnott *et al.* (1976)

[a]The first values in each column were calculated using the program by Rosenberg (Rosenberg *et al.*, 1976*b*), and the second by Hingerty (Hingerty, 1979).

attributed to the differences in conformations of the ribose sugars and the α angles in the two dinucleotides (table 2.2).

Loop structures for 3' terminal of RNA molecules have been proposed on the basis of the unusual structure of A2⁺pA3⁺ in ApApA and UpA1 in UpA. In both cases, the two phosphodiester bonds are in the *gauche*⁺ conformation. As pointed out by Kim *et al.* (1973) a g^+ γ angle causes a helical polynucleotide to fold back on itself. *Gauche*⁺ values for both β and γ cannot occur in the middle of a polynucleotide with 'rigid' nucleotide conformations because of sterically unfavourable phosphorus contacts.

The structure of these few oligomers has provided insight into the structural aspects of nucleic acid and also has allowed us to speculate a little about macromolecular structure. As it becomes possible to study larger fragments at higher resolution, the validity of these speculations can and will be tested.

ACKNOWLEDGMENTS

The help of Michael Dabrow with the figures and tables is gratefully acknowledged. This work was supported by grants from NIH GM 21589, CA 22780, CA 06927, RR 05539 and an appropriation from the Commonwealth of Pennsylvania.

REFERENCES

Arnott, S., Smith, P. J. C. and Chandrasekaran, R. (1976). In *Handbook of Biochemistry and Molecular Biology* (ed. C. D. Fasman), 3rd ed., vol. 2, section B, Chemical Rubber Co., Cleveland Ohio, p. 411.
Bugg, C. E., Thomas, J. M., Sundaralingam, M. and Rao, S. T. (1971). *Biopolymers*, **10**, 175.
Camerman, N., Fawcett, J. K. and Camerman, A. (1976). *J. molec. Biol.*, **107**, 601.
Drew, H. R., Dickerson, R. E. and Itakura, K. (1978). *J. molec. Biol.*, **125**, 535.
Hingerty, B. (1979). *Biopolymers*, **18**, 1901.
Hingerty, B., Subramanian, E., Stellman, S. D., Sato, T., Broyde, S. B. and Langridge, R. (1976). *Acta crystallogr.*, **B32**, 2998.
Kim, S. H., Berman, H. M., Seeman, N. C. and Newton, M. D. (1973). *Acta crystallogr.*, **B29**, 703.

Levitt, M. (1978). *Proc. natn. Acad. Sci. U.S.A.*, **75**, 640.

Rich, A., Davies, D. R., Crick, F. H. C. and Watson, J. D. (1961). *J. molec. Biol.*, **3**, 71.

Rosenberg, J. M., Seeman, N. C., Day, R. O. and Rich, A. (1976a). *J. molec. Biol.*, **104**, 145.

Rosenberg, J. M., Seeman, N. C., Day, R. O. and Rich, A. (1976b). *Biochem. biophys. Res. Commun.*, **69**, 979.

Rubin, J., Brennan, T. and Sundaralingam, M. (1972). *Biochemistry*, **11**, 3112.

Seeman, N. C., Sussman, J. L., Berman, H. M. and Kim, S. H. (1971). *Nature New Biology*, **233**, 90.

Seeman, N. C., Rosenberg, J. M., Suddath, F. L., Kim, J. J. P. and Rich, A. (1976). *J. molec. Biol.*, **104**, 109.

Suck, D., Manor, P. C. and Saenger, W. (1976). *Acta crystallogr.*, **B32**, 1727.

Sussman, J. L., Seeman, N. C., Kim, S. H. and Berman, H. M. (1972). *J. molec. Biol.*, **66**, 403.

Viswamitra, M. A., Kennard, O., Jones, P. G., Sheldrick, G. M., Salisbury, S., Falvello, L. and Shakked, Z. (1978). *Nature, Lond.*, **273**, 687.

Voet, D. and Rich, A. (1970). *Prog. Nucl. Acid Res. Mol. Biol.*, **10**, 153.

Wang, A. H.-J., Quigley, G. J., Kolpak, F. J., Crawford, J. L., van Boom, J. H., van der Marel, G. and Rich, A. (1979). *Nature, Lond.*, **282**, 680.

APPENDIX 2.1

The calculations of bend and twist angle are as follows.

1. *Z axis selection*

   Align the helical strands such that the portion of interest has the minor groove region towards the viewer. For base pair calculations the bases on the left-hand strand are used as reference and the bases on the right-hand strand are the target. The least squares plane of the reference base is called R and plane normal is N_r. Z axis is set to be parallel to N_r. The other base (target base) has least squares plane B and plane normal N_b. Both plane normals point up.

2. *Y axis selection*

   A plane (P) perpendicular to R and passing through two reference points, A_1 and A_2, is defined. A_1 is on reference ring which will be N1 if reference ring is a purine base or N3 for a pyrimidine base. A_2 will be the same type atom as A_1 on the target base. Y axis is defined to be parallel to the intersection line of planes P and R. Its direction will be further defined as positive in the direction of reference base to the target base.

3. *X axis selection*

   X axis is so selected that it will be perpendicular to both Y and Z axes. Its direction will point to the minor groove region if a right-handed system is adapted.

4. *Dihedral angle (D) calculation*

   If N_b decomposes to N_{bX}, N_{bY} and N_{bZ}, that is, the three component vectors along three axes, then the dihedral angle between two base planes can be expressed as

$$D = \cos^{-1} \frac{N_{bZ}}{(N^2_{bX} + N^2_{bY} + N^2_{bZ})^{1/2}}$$

5. *Bend angle (D_b) calculation*

$$D_b = \tan^{-1} \frac{N_{bY}}{N_{bZ}}$$

6. *Twist angle (D_t) calculation*

$$D_t = \tan^{-1} \frac{N_{bX}}{N_{bZ}}$$

Figure 2.7 shows the definition of the coordination system in a projection view and also shows some examples of base bending and twisting.

In the case of A-A pairing, the two reference points are A_1 = C5 of reference base and A_2 = N7 of target base. For U-U pairing, A_1 = C2 and A_2 = C4. These calculations also apply to single strand stacked bases. The strands are aligned so that they are viewed from the minor groove side parallel to the bases with the 3' base above the 5' base. The upper base is used as the reference and the lower one as the target. Both reference points are selected from reference base A_1 = C4 and A_2 = N1 if reference is a purine base, while A_1 = C6 and A_2 = N3 for pyrimidine base.

If the target base is considered in four aspects with respect to the reference base – major groove side, minor groove side, head (the side with Watson-Crick hydrogen bonding atoms) and tail (the opposite side of head) – then the results of bend and twist angles are interpreted as follows:

		Relationships of target to reference bases	
		Base pair	Stacked base
D_b	+	tail down	head down
	–	tail up	head up
D_t	+	major groove up	major groove up
	–	major groove down	major groove down

3

NMR studies of oligonucleotides in solution

Ramaswamy H. Sarma and M. M. Dhingra

INTRODUCTION

Crystal structures of oligonucleotides that have been determined to date are discussed in chapter 2. In the present chapter we take up their structure in aqueous solution as determined by NMR spectroscopy. Early NMR work on oligonucleotide conformations in solution was limited to investigations of base stacking and sugar ring pucker from base proton chemical shifts and anomeric proton spin coupling constants (Bangerter and Chan, 1969; Chan and Nelson, 1969; Hruska and Danyluk, 1968; Ts'o et al., 1969). In these earlier days the extreme complexity of the spectra as well as poor signal-to-noise ratios prevented a complete analysis. However, the development of high-frequency Fourier transform NMR systems with highly sophisticated interactive computer simulation capabilities and of selective deuteration capabilities, as well as hetero- and homonuclear decoupling in Fourier transform mode, have made it possible to solve this problem to some extent. As a result, the high resolution NMR spectra of many short, single stranded oligonucleotides (that is, up to trimer level) have been completely analysed (Cheng and Sarma, 1977a, b; Cheng, Dhingra and Sarma, 1978; Evans, Lee and Sarma, 1975; Evans and Sarma, 1976; Ezra, Lee, Kondo, Danyluk and Sarma, 1977; Lee, Evans and Sarma, 1975, 1976; Lee, Ezra, Kondo, Sarma and Danyluk, 1976) (figures 3.1, 3.2 and 3.3). Although increasing the number of nucleotidyl units beyond the trimer level makes the unambiguous assignment and complete analysis difficult, they do bring forth the prospects of studying the base-pairing interactions as well as the dynamics of helix-coil transition.

This chapter illustrates how NMR in its existing state of the art is being utilised to probe the conformational properties of single-stranded and base-paired short oligonucleotides in aqueous solution, but before examining these studies the section that follows looks briefly at the mononucleotides.

THE 3′ AND 5′ MONONUCLEOTIDES

Figure 3.1 shows the experimentally observed and computer simulated ^{1}H NMR spectra of a mononucleotide. The simulation provides accurate information about the coupling constants and chemical shifts which can be translated into conformational parameters using methodology described elsewhere (Dhingra and Sarma, 1979a; Sarma, 1979). The data summarised in table 3.1 have several general features.

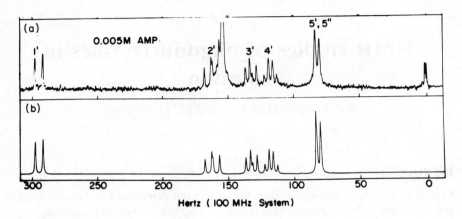

Figure 3.1   The sugar proton region of the 100 MHz ^{1}H NMR spectra of 5′AMP under conditions in which ^{31}P was decoupled, (a) observed and (b) computer simulated. The shifts are in hertz upfield from internal tetramethylammonium chloride.

Thus all 5′ mononucleotides irrespective of the nature of the base or the sugar ring display a great deal of tendency to orient the C4′—C5′ and C5′—O5′ bonds in the gg (ϵ = 60°) and $g'g'$ (δ = 180°) conformation. The sugar ring of the deoxyribo systems shows outspoken preference for the ^{2}E pucker in solution as in B-DNA. In the ribo series the ring pucker is sensitive to the constitution of the base, that is, the purines tend to shift to the left and the pyrimidine to the right of the equilibrium ^{2}E $\rightleftarrows$ ^{3}E. The deoxyribo and ribo systems also show important differences about the C3′—O3′ torsion. In the ribo series we have an equilibrium,

$$^3E\alpha^- \; \rightleftarrows \; ^2E\alpha^+$$

but in the deoxyribo series C3′—O3′ essentially occupies domains centred around 200°. In the above equilibrium α^- and α^+ are domains centred around 205° and 275°, respectively. NMR spectroscopy does not enable a precise determination of the sugar base torsion angle, χ_{CN}, in mononucleotides. However, the several qualitative methods described elsewhere (Dhingra and Sarma, 1979a; Sarma, 1979) clearly show that for common 5′ and 3′ pyrimidine monomers it lies in the *anti* domain and that within the *anti* domain ^{3}E sugar pucker is associated with a value of χ_{CN} smaller than that associated with a ^{2}E pucker. The purine systems probably exist as a blend of *anti* and *syn* conformers.

Table 3.1 Conformational parameters for ribo- and deoxyribomononucleotides in aqueous solution at 20 °C, pH = 5.0 (Cheng, Dhingra and Sarma, 1978; Cheng and Sarma, 1977; Evans, Lee and Sarma, 1975; Evans and Sarma, 1976; Ezra *et al.*, 1977; Lee, Evans and Sarma, 1975)

Nucleoside	Sugar ring % ^{3}E	Backbone % $gg(\epsilon = 60°)$	% $g'g'(\delta \simeq 180°)$	α (°)
5'AMP	40	77	72	—
5'GMP	37	71	71	—
5'UMP	46	89	75	—
5'CMP	55	77	75	—
3'AMP	31	82	—	240 ± 36
3'GMP	33	72	—	240 ± 37
3'UMP	56	68	—	240 ± 35
3'CMP	60	71	—	240 ± 35
d-5'AMP	28	63	70	—
d-5'GMP	32	63	67	—
d-5'TMP	32	59	74	—
d-5'CMP	33	70	75	—
d-3'AMP	23	71	—	202
d-3'GMP	26	57	—	202
d-3'TMP	33	57	—	199
d-3'CMP	35	58	—	202

The formation of a phosphodiester linkage between 3' and 5' monomers results in the shortest oligonucleotide – the dinucleoside monophosphate. A complete conformational analysis for a dimer involves the determination of (1) the two sugar base torsions – glycosidic torsions X_1 and X_2, (2) the mode of pucker of two sugar rings, (3) torsion angles α_1, β_1, γ_1, ϵ_1, (4) torsion of the free exocyclic CH_2OH group – ϵ' and δ' (see chapter 1 for nomenclature). In the case of a trimer, the situation is more complex, that is, three glycosidic torsions, mode of pucker of three sugar rings and twelve other torsion angles. The determination of these conformational features essentially involves assignment and analysis of NMR spectra, accurate extraction of chemical shifts and coupling constants and translation of these NMR parameters to conformational angles. The methodology is extremely time consuming, complex and difficult, sometimes bordering on the tortuous. For details see Dhingra and Sarma (1979a) and Sarma (1979).

STEREODYNAMICS OF RNA OLIGONUCLEOTIDES

Common ribodinucleoside monophosphates

Of the sixteen possible ribodinucleoside monophosphates, detailed conformational deduction has been made for fifteen (Evans, Lee and Sarma, 1975; Ezra *et al.*, 1977; Lee, Evans and Sarma, 1975; Lee *et al.*, 1976) by complete NMR analysis. The data is missing on GpG because of line broadening complications. A sample

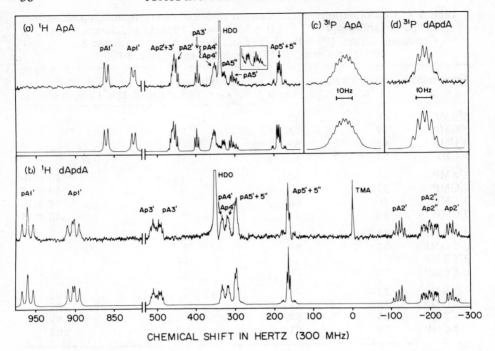

CHEMICAL SHIFT IN HERTZ (300 MHz)

Figure 3.2   (a), (b) 300 MHz ^{1}H NMR spectrum of the ribose region (a) ApA and (b) dApdA at 72° with chemical shifts expressed in hertz downfield from internal tetramethyl-ammonium chloride in D$_2$O. The simulations are presented beneath each NMR spectrum; (a) also contains an insert of the pA 5′ region at 68°. (c), (d) The 40.48 MHz ^{31}P FT NMR spectrum of (c) ApA at 40° and (d) dApdA at 27° along with simulations. Concentrations are 0.05 M, pD 7.4.

spectrum along with the simulation for a ribo dimer is shown in figure 3.2a. Complete conformational data for the dinucleoside monophosphates and the corresponding monomers are given in tables 3.2a and b.

The data unmistakably indicate that in all dimers, irrespective of the nature of the base and sequence, the C4′—C5′ and C5′—O5′ frame displays overwhelming preference for the *gg* and *g′g′* conformation. The ribose rings exist as a ^{2}E$\rightleftarrows$ ^{3}E equilibrium with bias for the ^{3}E pucker in most cases. Orientation about the C3′—O3′ bond is coupled to the ribose conformational equilibrium and the system exists with a bias for the ^{3}Eα^- coupled conformation. In most cases elevation of temperature (Ezra *et al.*, 1977) led to the following three experimental observations: (1) a reduction in the shielding between the adjacent bases indicating destacking (*vide infra*), (2) an increase in J1′2′ and a decrease in J3′4′ with the sum J1′2′ + J3′4′ and J2′3′ remaining constant, (3) a fine four bond coupling between H2′ and the phosphorus atom. These observations indicate that there is a conformational interconnectedness between C3′—O3′ torsion, sugar pucker and base–base interaction and led to the postulation of stereochemical domino effects (Dhingra and

Sarma, 1979b; Sarma, 1979; Sarma and Danyluk, 1977). We envision a coupled series of conformational events at the onset of stacking, made feasible by the swivel nature of the O3'—P—O5' bridge, that is, torsion about β and γ enables the bases to stack, causing a reduction in X_{CN} with an accompanying increase in ^3E populations, the latter in turn shifting the C3'—O3' to α^- domains.

The individual nucleotidyl units in the dimers differ in several key ways from the corresponding monomer conformations (table 3.2a, b). Specifically, the ribose ring becomes increasingly ^3E, and C4'—C5' and C5'—O5' become increasingly $gg\text{-}g'g'$, i.e. the molecule attempts to achieve conformational purity and identify with specific conformations upon oligomerisation.

The conformational properties of the individual segments of a dinucleoside monophosphate have been described, but to arrive at their intramolecular order and molecular topology it is also necessary to determine the sugar base torsions X_1 and X_2 as well as the phosphodiester torsions β and γ. All four angles are closely interrelated because changes in β and γ (under conditions in which α, δ and ϵ and sugar pucker occupy their preferred torsions) can bring the bases closer together to stack, which in turn will change X_1 and X_2.

No satisfactory methods exist to determine X_1, X_2, β and γ from coupling constants. An elaborate method based on the calculation of contribution to shielding

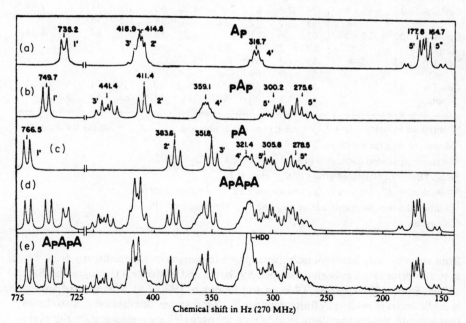

Figure 3.3 (a), (b), (c) Computer simulations of Ap-, -pAp- and -pA parts of ApApA. (d) Combination of the above three parts into one to produce (e) ApApA simulation. The 270 MHz ^{1}H experimental NMR spectrum of ApApA at 72°C, pD 7.0, 0.02 M. Note that the assignments of 1', 2' and 3' of Ap- and -pAp- are interchanged compared with the original report (Evans and Sarma, 1976).

Table 3.2 Population distribution of conformers in dinucleoside monophosphates and their components

(a)

			Dimer					Monomer[b]				
			Ribose ring[a]		Backbone[c]			Ribose ring[a]		Backbone[c]		
Nucleotide	Temp (°C)	% stacked[e]	%³E	K_{eq}^d	% gg	% g'g'	ΘPH[g]	%³E	K_{eq}^d	% gg	% g'g'	θPH[g]
ApA Ap- -pA	20	38 ± 2	58	1.4	79		± 37	31	0.5	82		± 36
			61	1.6	74	90		40	0.7	77	72	
ApA Ap- -pA	72	19 ± 4	46	0.9	73		± 33					
			48	0.9	64	73						
ApG Ap- -pG	20	25 ± 2	49	1.0	81		± 35	31	0.5	82		± 36
			51	1.0	75	82		37	0.6	71	71	
ApG Ap- -pG	80	16 ± 3	40	0.7	72		± 36					
			48	0.9	66	72						
GpA Gp- -pA	20	30 ± 2	55	1.2	72		± 36	33	0.5	72		± 37
			56	1.3	85	84		40	0.7	77	72	
UpU Up- -pU	20	8 ± 5	56	1.3	75		± 35	56	1.3	68		± 35
			53	1.1	83	77		46	0.8	89	75	
UpU Up- -pU	89	9 ± 2	52	1.1	71		± 35					
			52	1.1	77	73						
CpC Cp- -pC	20	35 ± 1	74	2.9	81		± 35	60	1.5	71		± 35
			69	2.2	92	84		55	1.2	77	75	
UpC Up- -pC	20	18 ± 2	62	1.6	73		± 35	56	1.3	68		± 35
			63	1.7	87	83		55	1.2	77	75	
UpC Up- -pC	80	0–4[f]	55	1.2	67		± 35					
			65	1.9	69	70						
CpU Cp- -pU	20	33 ± 2	66	1.9	78		± 33	60	1.5	71		± 35
			64	1.8	90	85		46	0.8	89	75	

[a]Computed by using $J_{1'2'} + J_{3'4'} = 9.5$ Hz for the dimers and $J_{1'2'} + J_{3'4'} = 9.3$ for the monomers.
[b]Monomer data for solutions at pD = 5.4.
[c]Rotamer equations used: $gg = (13.7 - \Sigma)/9.7$; $g'g' = (25 - \Sigma')/20.8$.
[d]K_{eq}, $^2E \rightleftarrows {}^3E$; estimated errors in K_{eq} are ± 0.2.
[e]% stacked = $(J_{3'4'}$ (dimer) $- J_{3'4'}$ (monomer)$)/(9.5 - J_{3'4'}$ (monomer)$)$.
[f]Computed from the magnitude of $J_{1'2'}$ and $J_{3'4'}$ Up-ribose moiety.
[g]In degrees.

from ring current, diamagnetic anisotrophy and paramagnetic anisotropy from x, y, z coordinates has been developed by Sarma and coworkers (Dhingra and Sarma, 1979a; Sarma, 1979; Sarma, Dhingra and Feldman, 1979a, b). This type of approach, in addition to providing definite data for χ_1, χ_2, β and γ, unmistakably also shows that in aqueous solution dinucleoside monophosphates are conformationally pluralistic. This pluralism results from the presence of a flexible conformational framework. While flexibility is allowed and alternate conformations are accessible, these molecules nevertheless attempt to achieve conformational identity by showing

Table 3.2 (cont)

(b)

Nucleotide	Temp (°C)	% stacked[e]	Dimer %³E	Keq[d]	%gg	%g'g'	θPH	Monomer[b] %³E	Keq[d]	%gg	%g'g'	θPH
			Ribose ring[a]		Backbone[c]			Ribose ring[a]		Backbone[c]		
ApU Ap--pU	20	34 ± 3	57	1.32	89		± 34	31	0.45	82		± 36
			59	1.44	95	85		46	0.85	89	75	
ApC Ap--pC	20	38 ± 2	64	1.78	85		± 33	31	0.45	82		± 36
			75	3.00	93	85		55	1.22	77	75	
ApC Ap--pC	80		44	0.79	69		± 36					
			61	1.56	69	70						
GpU Gp--pU	20	27 ± 5	49	0.92	80		± 34	33	0.49	72		± 37
			62	1.63	92	82		46	0.85	89	75	
GpC Gp--pC	20	45 ± 4	71	2.45	76		± 33	33	0.49	72		± 37
			79	3.76	96	89		55	1.22	77	75	
UpA Up--pA	20	15 ± 3	53	1.13	74		± 35	56	1.27	68		± 35
			51	1.04	79	80		40	0.67	77	72	
CpA Cp--pA	20	24 ± 2	71	2.45	79		± 35	60	1.50	71		± 35
			57	1.33	90	84		40	0.67	77	72	
UpG Up--pG	20	10 ± 5	54	1.17	75		± 35	56	1.27	68		± 35
			50	1.00	80	81		37	0.59	71	71	
CpG Cp--pG	20	25 ± 3	71	2.45	79		± 35	60	1.50	71		± 35
			57	1.33	85	85		37	0.59	81	71	
CpG Cp--pG	80		53	1.13	63		± 35					
			48	0.92	67	72						

[a]Computed by using $J_{1'2'} + J_{3'4'} = 9.5$ Hz for the dimers and $J_{1'2'} + J_{3'4'} = 9.3$ for monomers.
[b]Monomer data for solutions at pD = 5.4.
[c]Rotamer equations used, $gg = (13.7 - \Sigma)/9.7$; $g'g' = (25 - \Sigma')/20.8$.
[d]K_{eq}, $^2E \rightleftarrows {}^3E$; estimated errors in K_{eq} are ± 0.2.
[e]% stacked = $(J_{1'2'}(\text{monomer}) - J_{1'2'}(\text{dimer}))/J_{1'2'}(\text{monomer})$ or $(J_{3'4'}(\text{dimer}) - J_{3'4'}(\text{monomer}))/(9.5 - J_{3'4'}(\text{monomer}))$. The values are the average % stacked for the purine and pyrimidine fragments.

preferences, sometimes overwhelming, for certain orientations. The composition of the blend and the preferred intramolecular order are largely determined by the constitution and sequence which control the torsional movements about the various bonds in the molecular frame.

For example, UpA principally exists as a conformational blend of stacked, skewed, and extended arrays involving $g^- g^-$ ($\beta/\gamma \simeq 290°/290°$), $g^+ t$ ($\beta/\gamma \simeq 80°/180°$), $g^+ g^+$ ($\beta/\gamma \simeq 80°/80°$), tg^+ ($\beta/\gamma \simeq 180°/80°$) and tg^- ($\beta/\gamma \simeq 180°/290°$) conformers in which the $g^- g^-$ and $g^+ t$ predominates (Dhingra, Sarma, Geissner-Prettre and Pullman, 1978). On the other hand, UpU exists primarily in an extended array.

Modified dinucleoside monophosphates

A comparison of the conformational properties of common dinucleoside mono-phophates with those containing modified bases provides fundamental information about the influence of the nature of the base on the composition of the conformational blend. Both ApA and ϵApϵA (ϵA is an analogue of Y base in tRNA) have comparable populations (Dhingra, Sarma, Geissner-Prettre and Pullman, 1978) of g^-g^- arrays (figure 3.4). But the observation that the H5$'$ and H5$'$ of ϵAp- of ϵApϵA have undergone a shielding of 0.191 and 0.131 ppm, respectively, compared with 0.009 and 0.024 ppm in ApA, can be rationalised only on the ground that in ϵApϵA there is a significantly larger fraction of the g^+t conformer (figure 3.4) compared to ApA. In fact the lack of significant shielding of H5$'$ and H5$''$ of Ap-residue of ApA makes possible the conclusion that no significant amount of g^+t conformer is present in ApA. Comparison of the matched pair ϵApG and GpϵA gives insight in regard to the ability of ϵA to disturb the conformational equilibrium

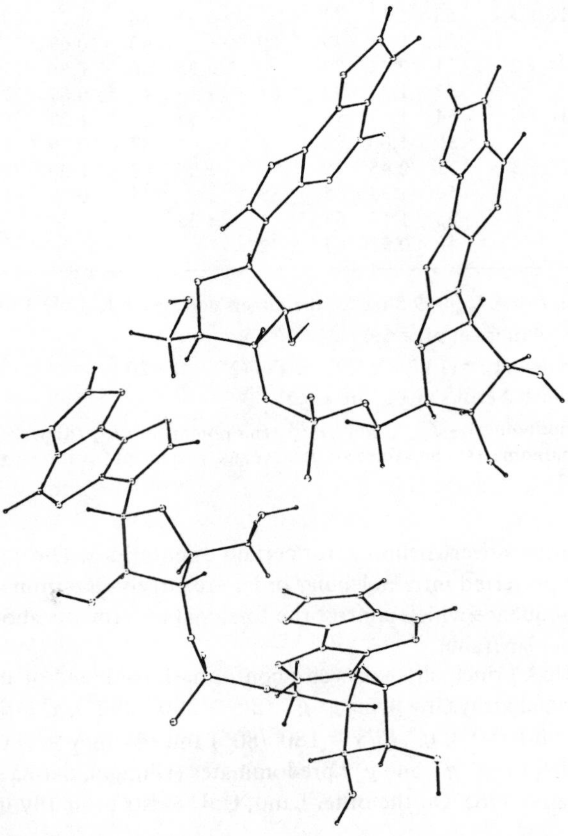

Figure 3.4 The g^-g^- (top right) and g^+t (bottom left) conformational arrays of ϵApϵA.

depending on whether it is located at the 3′ or 5′ end. When ϵA is moved from the 3′ to the 5′ end, the g^-g^- population increases and g^+t population decreases. In the case of mixed dimers, for example, in going from UpA to UpϵA, the H5′ and H5″ protons of the Np residue undergo substantial shielding (δH5′ = 0.44 ppm, δH5″ = 0.24 ppm), clearly indicating a substantial increase in g^+t arrays. Such large upfield shifts for the H5′ and H5″ of Np- have not been observed before.

Even modification of the base by methylation can have significant influence on the conformer distribution. NMR studes of m^1Apm^1A, m^1ApU and Upm^1A (Danyluk, Ainsworth and MacCross, 1978) indicate that methylation causes a shift in the ^{2}E $\rightleftarrows$ ^{3}E equilibrium to the left and increases the population of disordered unstacked structures, with little effect on α, δ and ϵ torsions. Comparison of the data for CpC and CmpC as well as ApA and AmpA (Cheng and Sarma, 1977a, b; Singh, Herbut, Lee and Sarma, 1976) indicates that methylation at the 2′ of the ribose can influence the conformational properties of these molecules. Thus 2′—O-methylation of the Cp- of CpC causes (1) a reduction in the magnitude of X_1, (2) an increase in the population of ^{3}E pucker at Cp-, (3) perturbations of β and γ. In the case of ApA, the significant effect of 2′—O methylation of Ap- results in destacking, a shift of ^{2}E $\rightleftarrows$ ^{3}E to the left for Ap- and a change in α. The implications of these findings with respect to biologically functional polynucleotides are discussed on p. 45.

Ribotrinucleoside diphosphates: looped-out and bulged configurations

The conformational deductions of ribodinucleoside monophosphates have shown that the spatial configuration properties of RNA oligomers are strongly influenced by chain length and sequence, with the monomeric units undergoing interdependent sterodynamical changes as they become integrated into the framework of ribodimers. A study of the trinucleoside monophosphates will indicate how these conformational changes are transmitted and propagated beyond the dimer level.

Theoretically, there are 64 possible ribotrinucleoside diphosphates. A complete analysis of an ^{1}H NMR spectra of a ribotrimer is a formidable task. However, with the help of selective deuteration, homo and heteronuclear decouplings and high field NMR studies, so far complete analysis of the 270 MHz ^{1}H NMR spectra of three ribotrimers have been completed (Dhingra and Sarma, 1979a, b; Sarma, 1979; Sarma $et\ al.$, to be published).

Figure 3.3 illustrates the ^{1}H NMR spectra of ApApA along with its complete simulation and assignments. Analysis of the NMR data from the trimers ApApA, CpCpA and ApCpC indicate that the dominant spatial array in the conformational blend is what is expected from the trend seen in progressing from monomer → dimer → trimer, that is in each residue the sugar pucker is ^{3}E, $\alpha \simeq 205°$, $\beta\gamma = g^-g^-$, $\delta \simeq 60°$, $\epsilon = 60°$ and sugar base torsion $anti$. Such a spatial configuration (Sarma $et\ al.$, to be published) for CpCpA is illustrated in figure 3.5.

The data reveal that $in\ the\ progression\ from\ monomer\ to\ dimer\ and\ dimer\ to\ trimer\ the\ major\ conformational\ change\ takes\ place\ at\ the\ first\ step,\ in\ other\ words,\ from\ dimer\ to\ trimer\ the\ magnitude\ of\ changes\ are\ small$. The conformational free-

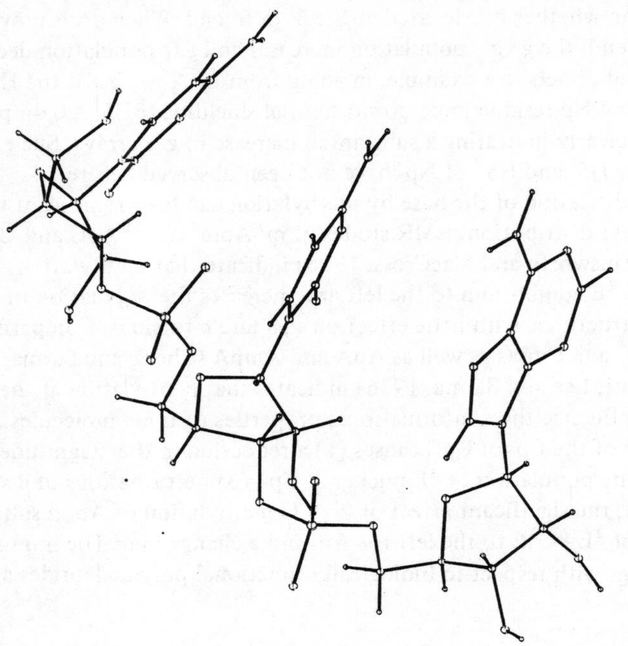

Figure 3.5   The preferred spatial configurations of CpCpA in aqueous solution.

dom and flexibility in the trimer enables it to assume a variety of spatial configurations as the less dominant ones. For example, in the cases of CpCpA and ApCpC, evidence has been found (Sarma *et al.*, to be published) for the presence of unusual looped-out spatial configurations in which the central cytidine unit -pCp- is bulged out, enabling stacking interactions between the terminal adenine and cytosine bases. Some of the details of these will be discussed in connection with deoxyribotrinucleoside diphosphates (p. 48).

Further, a comparison of the spatial configuration data for ApCpC and CpCpA provides insights regarding the effect of sequence on local conformational flexibility. For example, in CpCpA the population of $\epsilon_2 = 60°$ conformers is noticeably less compared to CpA; movement of adenine from the 5′ end (ApCpC) to the 3′ end (CpCpA) shifts the $^2E \rightleftarrows {}^3E$ equilibrium towards ^2E to the extent that the sugar ring of -pA of CpCpA displays equal proclivity for ^2E and ^3E conformations. This information, along with the observation (*vide supra*) that in aqueous solution ^2E is associated with $\alpha \simeq 275°$ and ^3E with $\alpha \simeq 205°$, and in the *anti* domain χ_{CN} is larger for the ^2E systems compared to the ^3E, clearly indicate that, in addition to the spatial array shown in figure 3.5, conformers in which the geometric details of the adenosine moiety display considerable variation from other residues make noticeable contribution to the conformational blend of CpCpA, the acceptor end of tRNA.

RIBOMONONUCLEOTIDE CONFORMATION APPLIED TO RIBOPOLY—NUCLEOTIDES

From the extensive examination of solid-state data Sundaralingam (1973, 1975) has concluded that nucleotides are considerably more rigid than nucleosides and that nucleotides essentially maintain their isolated conformations when they become part of a polynucleotide. This is the concept of a rigid nucleotide. The concept also states that the nucleic acids may achieve conformational changes by torsional variations about $O3'$—P and P—$O5'$ bonds.

Aqueous solution data on purine ribonucleosides, and the corresponding 3' and 5' mononucleotides as well as 3', 5' nucleoside diphosphates clearly reveal that the backbone conformations of ribonucleosides are as flexible as those of the corresponding nucleotides (Evans and Sarma, 1976). Examination of the ^{1}H NMR parameters in adenosine, 3'-AMP, 5'-AMP, 3',5'-ADP, ApA, ApApA, and poly A under various conditions reveals that the conformational destiny of mononucleotides as they become part of polynucleotides is ordained by their isolated conformations only when significant amounts of the polymer are in the destacked state. The con-

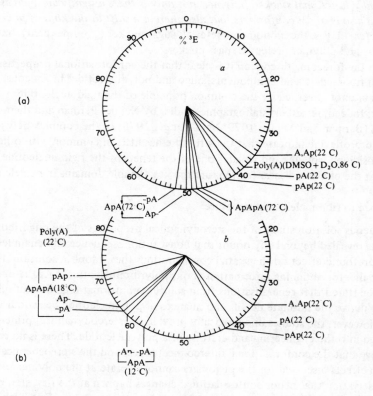

Figure 3.6 Computed percentage populations of the ^{3}E ribose conformers in adenine nucleotide systems under (a) destacked and (b) stacked conditions (Evans and Sarma, 1976).

formational conservation breaks down as soon as the mononucleotides become integrated into the backbone framework of a biologically functional base-stacked polynucleotide. This is dramatically illustrated in figure 3.6 where the sugar conformation data are presented for adenosine, $3'$ AMP, $5'$ AMP, $3'5'$ ADP, ApA, ApApA and poly A under conditions of high base stacking (low temperature) and low base stacking (high temperature). Base-stacking interactions have a profound impact on the conformational properties of ribonucleic acids; in aqueous solution their major effect is to diminish conformational freedom. Comparison of the computed percentage populations of ^3E conformers for stacked poly A, ApApA, and ApA with that of the component monomers (figure 3.6) reveals that the ribose ring in the mostly stacked oligomers and polymer displays a conformational preference dramatically different from that of the monomeric components. The ^3E populations in the monomers are about 40 per cent, while the corresponding populations for ApA, ApApA and poly A in conditions which favour stacking lie in the range of 60–80 per cent. The higher end of the range is preferred by the central -pAp- unit of ApApA and the nucleotidyl units in poly A. It should be noted that *as the monomeric components become integrated into the framework of a biologically functional stacked polymer, not only is there a significant increase in the populations of ^3E conformers, but also there is a shift in the kind of pucker they prefer*, that is, the monomers prefer ^2E sugar pucker ($\simeq$ 60 per cent), but the oligomers and polymer prefer ^3E sugar pucker.

From the foregoing discussion it is clear that the conformational properties of common ribonucleic acid components, oligo and polynucleotides in aqueous solution are not governed by the common principle of the rigid nucleotide concept. Indeed, recent crystallographic studies by Neidle, Berman and their co-workers (Berman and Neidle, 1979; Neidle *et al.*, 1978) of the complex between ApA and proflavine indicate that the structure has little in common with other dinucleoside phosphates and violates one of the tenets of the rigid nucleotide concept that the phosphodiester torsions are the key flexible domains in nucleic acids.

Relevance to ribonucleic acid structure

The aqueous solution study of the sterodynamical properties of the oligoribonucleotides has revealed beyond any doubt that constitution, sequence and chain length have a profound effect on the spatial configurations they adopt in solution. Is this of any value for obtaining information about polyribonucleotides? At the outset it should be stated that generalisations from solution conformational data of oligomers to derive the intimate details and nuances of polynucleotide behaviour are not valid. However, the studies unequivocally provide the stereodynamical principles which go into the engineering and crafting of a polynucleotide. There is no reason to believe that the conformational interconnectedness and the stereochemical domino effects discussed for the oligomers cannot operate at the polymer level. Our observation that major conformational changes happen at the first step in oligomerisation, that is, from monomer to dimer, and that at the level of a trimer there is considerable conformational flexibility, suggests that we cannot expect in aqueous solution a polynucleotide with single rigid spatial configuration and that it must possess a great deal of flexibility. Recent examination of the anatomy of tRNA *vis-à-vis* polynucleotide flexibility by Rich, Quigley and Wang (1979) supports this thesis; they found the molecule adopting unusual conformations associated with chain extension, changes in direction of the polynucleotide chain or with the accommodation of other bases which intercalate into the chains. The observation that in aqueous solution the oligonucleotide CpCpA is structured in such a way that it is capable of adopting a variety of spatial configurations with considerable flexibility for the terminal adenine nucleotide unit suggests that it may be so for the acceptor end of tRNA. Examination of the four reported crystal structures of tRNA, two from orthorhombic (Sussman and Kim, 1976; Quigley *et al.*, 1975) and two from monoclinic (Ladner *et al.*, 1976; Stout *et al.*, 1976), indicate considerable variation in the structure of residues 74, 75 and 76 which could be a manifestation of the intrinsic flexibility of the CpCpA oligomer. Such conformational variability may be implicated in recognition of aminoacyl tRNA synthetase. The observation of the effect of methylation of the base or the sugar on the conformational properties of dimers suggests that one of the roles of methylation may be to induce local conformational changes in the polymer. It is striking that residue 58 in tRNA^{Phe} is N^1 methylated adenine nucleotide and is the linking nucleotidyl residue between TψC arm (stem) and TψC loop regions where the ordered stacked structure of the stem region alters into a more open loop conformation and that N^1 methylated adenylate dimers themselves have a disordered spatial configuration. Study of the conformational effects of 2′—O-methylation of ribose in dimers and 7-methylation of guanosine have been of great value to arrive at the spatial configuration of the bizarre 5′ terminus of mammalian messenger. A structure in which the 7-methyl guanosine intercalates into a chain extended AmpA, the extension of the chain being caused by 2′—O-methylation, has been arrived at recently for the 5′ terminus of mRNA (Kim and Sarma, 1978).

STEREODYNAMICS OF DNA OLIGONUCLEOTIDES

From the extensive NMR studies of ribo-oligonucleotides reported above we have obtained information about their conformational features, properties and dynamics. An important question is whether the simple substitution of a 2′OH by hydrogen in the sugar ring will have far-reaching repercussions with respect to the three-dimensional spatial configuration and whether such conformational effects may enable us to understand the profound differences that exist in the physico-chemical and biological properties between ribo and deoxyribonucleic acid structures. This is attempted below.

Deoxyribodinucleoside monophosphates

Of the sixteen possible deoxyribodinucleoside monophosphates, detailed conformational deductions have been made for fifteen by complete analysis of their NMR spectra (Cheng and Sarma, 1977*a, b*). A sample spectrum along with the simulation

Table 3.3 Population distribution of conformers in deoxyribonucleoside monophosphates and their components

| Nucleotide | Temp, °C | Dimer | | | | | | Monomer[d] | | | | |
| | | Sugar ring[a] | | | Backbone[b] | | | Sugar ring[a] | | Backbone[b] | | |
		% ³E	K_{eq}[c]		% gg	% g'g'	θPH, deg	% ³E	K_{eq}[c]	% gg	% g'g'	θPH, deg
d-ApA dAp-	27	22	0.3		73		± 45	23	0.3	71		± 38
-pdA		37	0.6		87	85		28	0.4	63	70	
d-ApA dAp-	72	20	0.3		59		± 43					
-pdA		37	0.6		76	76						
d-GpG dGp-	80	28	0.4		57		± 45					
-pdG		39	0.6		90	86						
d-ApG dAp-	27	19	0.2		65		± 45	23	0.3	71		± 38
-pdG		31	0.4		94	86		32	0.5	63	67	
d-ApG dAp-	72	26	0.4		57		± 44					
-pdG		37	0.6		76	76						
d-GpA dGp-	20	33	0.5		69		± 40	26	0.4	57		± 38
-pdA		37	0.6		84	86		28	0.4	63	70	
-d-GpA dGp-	80	28	0.4		58		± 42					
-pdA		32	0.5		53	74						
dCpC dCp-	20	33	0.5		67		± 40	35	0.5	58		± 38
-pdC		32	0.5		75	76		33	0.5	70	75	
dCpC dCp-	80	33	0.5		67		± 40					
-pdC		38	0.6		61	69						
d-TpT dTp-	18	30	0.4		63		± 41	33	0.5	57		± 41
-pdT		37	0.6		74	82		32	0.5	59	74	
d-TpT dTp-	65	33	0.5		49		± 39					
-pdT		34	0.5		64	77						
d-CpT dCp	20	33	0.5		63		± 37	35	0.5	58		± 38
-pdT		32	0.5		71	74		32	0.5	59	74	
d-CpT dCp-	80	31	0.4		68		± 37					
-pdT		37	0.6		73	76						
d-TpC dTp-	20	27	0.4		56		± 38	33	0.5	57		± 41
-pdC		32	0.5		79	77		33	0.5	70	75	
-TpC dTp-	80	32	0.5		56		± 38					
-pdC		34	0.5		69	71						
d-ApC dAp-	27	26	0.4		71		± 41	23	0.3	71		± 38
-pdC		37	0.6		63	82		33	0.5	70	75	
d-ApC dAp-	80	31	0.5		61		± 41					
-pdC		37	0.6		53	72						
d-ApT dAp-	27	32	0.5		71		± 44	23	0.3	71		± 38
-pdT		41	0.7		86	89		32	0.5	59	74	
d-ApT dAp-	80	26	0.4		65		± 39					
-pdT		49	1.0		42	72						
d-GpC dGp-	20	31	0.5		65		± 40	26	0.4	57		± 38
-pdC		37	0.6		73	82		33	0.5	70	75	
d-GpC dGp-	80	31	0.5		63		± 41					
-pdC		37	0.6		63	73						

Table 3.3 (cont)

		Dimer					Monomer[d]				
		Sugar ring[a]			Backbone[b]			Sugar ring[a]		Backbone[b]	
Nucleotide	Temp, °C	% ³E	K_{eq}^{c}	% gg	% g'g'	θPH, deg	% ³E	K_{eq}^{c}	% gg	% g'g'	θPH, deg
d-GpT dGp-	20	30	0.4	55		± 39	26	0.4	57		± 38
-pdT		44	0.8	81	82		32	0.5	59	74	
d-GpT dGp-	80	33	0.5	55		± 41					
-pdT		31	0.5	81	82						
d-CpA dCp-	20	28	0.4	58		± 40	35	0.5	58		± 38
-pdA		36	0.6	90	86		28	0.4	63	70	
d-CpA dCp-	80	28	0.4	58		± 39					
-pdA		36	0.6	63	75						
d-CpG dCp-	20	33	0.5	55		± 40	35	0.5	58		± 38
-pdG		37	0.6	75	86		32	0.5	63	67	
d-CpG dCp-	80	25	0.3	65		± 44					
-pdG		37	0.6	56	75						
d-TpA dTp-	20	24	0.3	55		± 37	33	0.5	57		± 41
-pdA		36	0.6	90	86		28	0.4	63	70	
d-TpA dTp-	80	33	0.5	63		± 40					
-pdA		36	0.6	61	80						
d-TpG dTp-	20	24	0.3	55		± 37	33	0.5	57		± 41
-pdG		35	0.5	75	86		32	0.5	63	67	
d-TpG dTp-	80	24	0.3	55		± 35					
-pdG		29	0.4	79	88						

[a]Computed by using $J_{1'2'} + J_{3'4'} = 10.8$ Hz for the dimers and $J_{1'2'} + J_{3'4'}$ for the monomers.
[b]Rotamer equations used: $gg = (13.7 - \Sigma)/9.7$; $g'g' = (25 - \Sigma')/20.8$.
[c]K_{eq} $^2E \rightleftarrows {}^3E$.
[d]Monomer data for solutions at pD 5.4.

for a deoxyribodimer is shown in figure 3.2b. The conformational parameters for the dimers and monomers are given in table 3.3. The data indicate that the deoxy dimers exist as a conformational blend in which certain spatial arrays predominate. The pentose ring exists as a $^2E \rightleftarrows {}^3E$ equilibrium with significant preference for the ^2E pucker in all cases, irrespective of the nature of the base and sequence. The C4'—C5' and C5'—C5' bonds form a conformational network in which $\delta \simeq 180°$ and $\epsilon \simeq 60°$ are preferred. The C3'—O3' bond occupies a domain in which $\alpha \simeq 200°$. Except for d-TpT, d-TpC and d-CpT the dimers exist significantly in stacked arrays in which $\beta\gamma$ is in the g^-g^- domain. Elevation of temperature and consequent destacking have only minor effects on pentose conformation and that about α. However, it leads consistently to changes of β from g^- to t domains with little effect on γ. The dimers d-TpT and d-TpC display tg^- conformation for the phosphodiester bonds. Oligomerisation of monomer to dimer results in significant increase in the population of *gg* and *g'g'* conformers about the internucleotide C4'—C5' and C5'—O5' bonds with no noticeably important changes in other conformational parameters.

Deoxyribotrinucleoside phosphates: Bulged and looped-out configurations

Complete analysis of the ^{1}H NMR spectra of two deoxyribotrinucleoside diphosphates (d-TpTpT and d-TpTpC) at 20° and 80°C have been reported so far (Cheng, Dhingra and Sarma, 1978). The observed NMR parameters indicate that the conformational properties of the trimers are very similar to those of constituent dimers, that is in the deoxy systems the dimers conserve their intrinsic conformational features when they are incorporated into the oligomers. The preferred spatial configuration of d-TpTpT is shown in figure 3.7a.

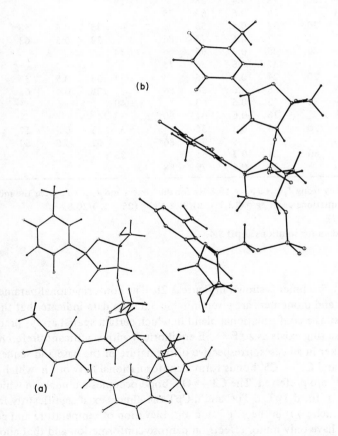

Figure 3.7 (a) The preferred spatial configuration of d-TpTpT in aqueous solution. In each of the individual nucleotidyl units $\chi_{cn} = anti$ and sugar pucker is ^{2}E; in addition ϵ', ϵ_1, $\epsilon_2 = 60°$, δ_1, $\delta_2 = 180°$, α_1, $\alpha_2 = 199°$, $\beta_1 \gamma_1$ and $\beta_2 \gamma_2 = tg^-$. (b) The preferred spatial configuration of d-TpTpA in aqueous solution. Except for $\beta_2 \gamma_2 = g^- g^-$, all angles are the same as in (a).

The conclusion that the trimers d-TpTpT and d-TpTpC essentially maintain the isolated conformations of the corresponding dimers agrees with similar conclusions by Kan, Barrett and Ts'o (1973) about d-ApTpT and d-TpTpA from an incomplete analysis of their NMR spectra. However Kan *et al.* conclude that the dimer segments of d-ApTpT and d-TpTpA significantly populate in the naturally preferred right-handed arrangement ($\beta_1/\gamma_1 \simeq 290°/290°$ and $\beta_2/\gamma_2 \simeq 290°/290°$). Recent extensive ^{1}H NMR studies (Cheng and Sarma, 1977a) of d-TpT, d-ApT and d-TpA have shown that d-TpT prefers the tg^- domain ($\beta/\gamma \simeq 180°/290°$) and d-ApT and d-TpA the g^-g^- ($\beta/\gamma \simeq 290°/290°$) domain. Based on this new information the conclusion should be that the phosphodiester bonds of the d-TpT segments of d-TpTpA and d-ApTpT preferentially exist in tg^- orientation and that of the d-TpA and d-ApT segments prefer the traditional g^-g^- domain (figure 3.7b). In the conformation shown in figure 3.7b the magnitude of cylindrical coordinates are such that H6, H1$'$ and the methyl protons of dTp-residue will experience no shielding from the adenine moiety. This means that the actually observed distant shielding of the terminal dTp-residue of d-TpTpA by the terminal adenine (Kan, Barrett and Ts'o, 1973) does not originate from the preferred conformation of the trimer (figure 3.7b). Kan *et al.* have suggested that the distant shielding may originate from right-handed stacks. In such stacks β_1 γ_1 and β_2 γ_2 occupy g^-g^- domains, and the bases are stacked, and it is clear from the cylindrical coordinates (Cheng, Dhingra and Sarma, 1978) z, ρ_5 and ρ_6, the bases of the end nucleotidyl units are too far apart to have any mutual shielding effects. Hence the observed distant shieldings cannot originate from conformations in which both the phosphodiester bonds occupy g^-g^- domains, even though such conformers may make minor contribution to the conformational blend.

We believe that the distant shieldings in d-TpTpA and d-ApTpT originate from certain minor conformers which contribute to the conformational blend. Because the molecule is flexible it may adopt different conformations in solution. The proclivity of base pairs such as purines–purines and pyrimidine–purines to interact and stack provide the driving force for the two end nucleotidyl units to undergo torsional motion about the phosphodiester bonds so that the bases can interact. In order to determine the spatial configuration of a trimer in which the two end nucleotidyl bases can interact and cause mutual shifts, we have examined a reasonable number of β_1, γ_1 and β_2, γ_2 combinations. The combinations used are $|g^+g^-$, $g^+g^+|$, $|g^+g^-, g^+g^-|g^+g^+, g^-g^+|$, $|g^+g^-, g^-g^+|$ and $|g^+g^+, g^+g^+|$. The perspectives of some of these combinations are shown in figure 3.8a, b, c, d. As is clear from figure 3.8a, b in the $|g^+g^-, g^+g^-|$ and $|g^+g^-, g^+g^+|$ combinations, the bases of the end nucleotidyl units can be parallel but the value of z is too large to account for any distant shielding effects. In the combinations $|g^+g^-, g^-g^+|$ and $|g^+g^+, g^-g^+|$ (figure 3.8c, d) the bases cannot be parallel for reasonable values of χ_{cn}, and the distance between the base planes are too large to account for any distant shielding effects. The structure in which the phosphodiester bonds assume $|g^+g^+, g^+g^+|$ conformation generates a spatial array in which the end nucleotidyl bases are approximately parallel and the value of z is reasonable to induce the ring current shifts. Hence in the cases of d-TpTpA and d-ApTpT the bases of the end nucleotidyl units

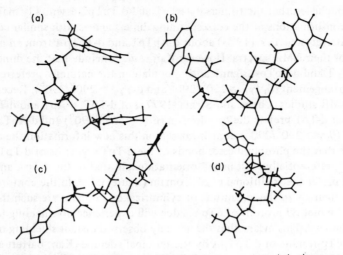

Figure 3.8 Perspectives of d-TpTpA in which $\beta_1\gamma_1$ and $\beta_2\gamma_2$ are in (a) $|g^+g^+, g^+g^-|$, (b) $|g^+g^-,$ $g^+g^+|$, (c) $|g^+g^+, g^-g^+|$, and (d) $|g^+g^-, g^-g^+|$.

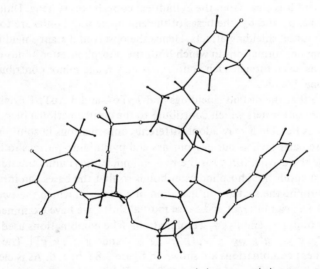

Figure 3.9 Perspective of d-TpTpA in which $\beta_1\gamma_1 = g^+g^+$, $\beta_2\gamma_2 = g^+g^+$. Rest of the details for the backbone as in figure 3.7.

can interact if $\beta_1\,\gamma_1$ occupies g^+g^+ and $\beta_2\,\gamma_2$ occupies g^+g^+ domains. A perspective of such a conformation for d-TpTpA is shown in figure 3.9. It is clear from this figure and the corresponding cylindrical coordinates (Cheng, Dhingra and Sarma, 1978) that in this structure the adenine ring will shield the H1′, H6 and methyl protons of dTp- residue. The projected shieldings are larger than the observed ones because conformation in figure 3.9 is only a minor contributor.

Spatial configurations in which $\beta\gamma$ occupies the g^+g^+ domains have been reported for most of the ribo and deoxyribodinucleoside monophosphates in solution (Ezra *et al.*, 1977; Lee *et al.*, 1976) as well as in a few crystal structure studies (Seeman, Day and Rich, 1975; Seeman *et al.*, 1971, 1976; Suck *et al.*, 1973). Even from simple qualitative arguments the existence of conformers such as those in figure 3.9 for d-TpTpA and d-ApTpT could be predicted. It is known that there is very little base–base interaction in d-TpT and that d-TpA and d-ApT significantly populate in stacked arrays (Cheng and Sarma, 1977*a*). Hence it is entirely reasonable to expect the dTp- residue in d-TpTpA to attempt to interact with the -dpA segment of d-TpTpA to generate a 'counterfeit d-TpA' structure. The flexibility of the phosphodiester bonds indeed allows such an interaction, that is, the inherent flexibility of the phosphodiester bonds in structures like d-TpTpA enables a competition between the middle and the end thymidyl residues to interact with the terminal adenine. Because of the poor ring current fields of the pyrimidine bases, the NMR data on d-TpTpT and d-TpTpC (Cheng, Dhingra and Sarma, 1978) are of little use to determine whether there is interaction between the end nucleotidyl units in these systems.

COMPARISON BETWEEN RNA AND DNA SINGLE-STRANDED OLIGOMERS AND RELEVANCE TO BIOLOGICAL STRUCTURES

In the deoxyribo-3' and -5' mononucleotides, as well as in the dimers, irrespective of the nature of the base and sequence, the pentose ring shows a clear preference for the ^2E conformation. This is the case in every naturally occurring constituent of deoxyribonucleic acids. Furthermore, the conformational distribution of the pentose does not show any meaningful sensitivity to such strong intramolecular perturbations such as stacking/destacking interactions. Thus it is reasonable to conclude that the conformation of the pentose moieties in single-stranded hetero and homodeoxyribopolynucleotides in aqueous solution resembles that in their constituents and that the pentose moieties in them will show a great deal of preference for the ^2E mode of pucker. We realise that in crystals both ^2E and ^3E DNAs have been reported. We do not have any evidence that there is any chance that a given sequence of single-stranded DNA in aqueous solution could populate predominantly in the ^3E form.

This is in sharp contrast to the behaviour of ribonucleic acid structures where in aqueous solution the kind of sugar pucker a system prefers is determined by the nature of the base and stacking interactions. For example, the purine ribomononucleotides prefer ^2E pucker, the pyrimidines ^3E pucker. Stacking interactions cause a shift of $^2E \rightarrow {}^3E$ pucker and under conditions in which significant amounts of the polymer exist destacked the preferred pucker for the sugar is ^2E.

In the ribo series it has been shown that torsional variation about C3'—O3' is coupled to the ribose conformation and governed by the equilibrium (Cheng and

Sarma, 1977*a*; Ezra *et al.*, 1977; Lee *et al.*, 1976).

$$^{3}E\alpha^{-} \rightleftarrows {}^{2}E\alpha^{+}$$

In the deoxyribo series the favoured conformational coupling is between ^{2}E and α^{-}.

The majority of ribo and deoxyribodinucleoside monophosphates prefer to exist in stacked arrays in which $\beta\gamma$ lie in the $g^{-}g^{-}$ domain. Because of the difference in the sugar pucker it is obligatory that there should be local differences between the two classes within the $g^{-}g^{-}$ conformation space.

An important observation is that temperature-induced unstacking causes changes in β from g^{-} to t with little effect on γ. This suggests that the main torsional event at the onset of unwinding double helical DNA and RNA is that about O3$'$—P. In both ribo and deoxyribo series changes in $\beta\gamma$ are accompanied by destacking and χ_{1} and χ_{2} changes. Such χ_{1} χ_{2} changes alter the mode of sugar pucker and torsion about C3$'$—O3$'$ in the ribose series, but no such changes are noticeable for the deoxy systems. This suggests that polyribonucleotides are engineered in such a way that their conformations are very sensitive to minor perturbations and that it is possible that they are capable of fulfilling their multifunctional biological roles as tRNA, mRNA and rRNA because of this built-in potentiality in their constitution for conformational versatility and pluralism. Even though the solution data suggest that deoxyribopolynucleotides may be conformationally less versatile than the ribopolymers, crystal data indicate that the deoxys exhibit greater flexibility in the sugar pucker as in the *A* and *B* forms of DNA. They are permitted both ^{2}E and ^{3}E pucker in the $\beta\gamma = g^{-}g^{-}$ domain. The former is sterically disflavoured in the RNAs. Consequently in crystalline tRNA, for example, only a few of the approximately 80 residues have sugar pucker ^{2}E and the reason for this is elaborated by Rich, Quigley and Wang (1979).

So far discussions have been limited to short oligonucleotides whose NMR spectra have been completely analysed. Some information is available from incomplete analysis of higher single stranded oligomers. Ts'o and coworkers (Borer, Kan and Ts'o, 1975; Kan, Borer and Ts'o, 1975; Kan *et al.*, 1975) have carried out the assignment of base proton and H1$'$ region in r-ApApCpCpUpU and r-GmpApApYpApψp by a systematic incremental procedure, that is by probing the effect of added nucleotidyl units. The authors have interpreted their data in terms of a spatial configuration in which the purine bases have a maximal extent of base–base overlap with their neighbours in sequence and that this manifests in a zigzag balcony structure for the -ApYpA fragment.

BASE-PAIRED OLIGONUCLEOTIDES

While the application of NMR methods has yielded a wealth of information about the geometry and dynamics of single-stranded oligonucleotides, their application to base paired oligonucleotides has not yielded comparable information. This is primarily because of the difficulty in analysing complex spectra from the duplexes from large oligomers. So far there has been only a single complete analysis of the spectra of a duplex and this is for a self-complementary duplex of an analogue of

ApU in which the adenine and uracil were chemically linked to the sugar moieties (Dhingra and Sarma, 1979a, b; Sarma, 1979; Sarma, Dhingra and Feldman, 1979) at a χ_{cn} of approximately 120°. This study provided the first experimental support for vertically stabilised double helices in which the base planes are parallel to the helical axis and the backbone is right handed for high *anti* polynucleotides, as projected by Olson (1977) from theoretical calculations. Most of the NMR studies of the duplexes have been concerned with effect of composition, sequence and chain length and helix coil transition. From these some qualitative information about their spatial configuration has emerged.

Base-paired oligomers of the RNA family

Krugh and colleagues (Krugh, Laing and Young, 1976; Young and Krugh, 1975) have monitored the shift of the amino protons in a number of self-complementary ribodinucleoside monophosphates CpG, GpC, UpA and the complementary mixture GpU + ApC as a function of concentration at low temperatures (figure 3.10). They observed a large downfield shift for the guanine amino and one of the cytosine amino protons as a function of increasing concentrations in CpG and GpC at 1 °C and this led them to conclude that self-complementary ribo dimers form Watson-Crick base-paired duplexes. In the complementary mixture GpU + ApC, the effect of hydrogen bonding on the shift of guanine amino protons was observed, but the adenine amino protons were not affected. The data led them to conclude that the GpU̲/C̲pA duplex̲ is̲ much less stable than either of the self-complementary duplexes GpC/CpG or CpG/GpC. This is an indication of the effect of the number of GC pairs on stability. The self-complementary UpA was found not to associate, again suggesting the stability differences between GC and AU pairs.

The base protons were found to shift upfield with increasing concentration in the range of 1 to 100 mM (Krugh, Laing and Young, 1976; Young and Krugh, 1975) (figure 3.10). They observed that the magnitude of the changes in H5 of the cytidine residue are significantly larger than those predicted for various double helical conformations of nucleic acids, possibly due to intermolecular aggregation. Such complications prevent the extraction of detailed geometric information concerning the conformation of these hydrogen bonded complexes.

Arter *et al.* (1974) observed only two resonances at 13.18 and 1.245 ppm* due to GC pairs in the self-complementary r-CpCpGpG, suggesting the formation of a symmetrical double helical fragment. Direct evidence for the formation of the duplex was provided by the observation of a melting curve (T_m = 51 ± 2 °C) for the base protons H8 and H2. In order to obtain some information about the geometry of the complex, they calculated the shielding (Young and Krugh, 1975) in a variety of duplex configurations and compared them with the observed shifts. The authors report that the observed shifts are consistent with those calculated for RNA-11.

*All shifts are with respect to DSS.

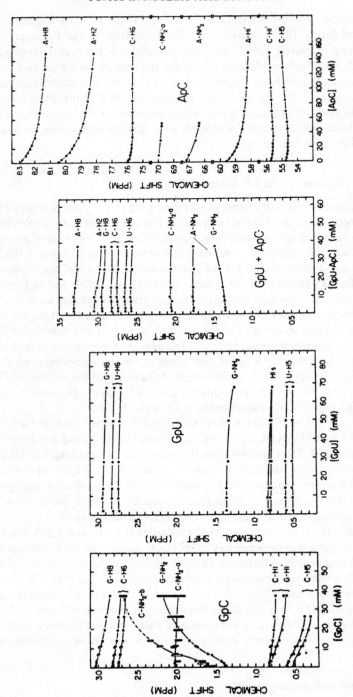

Figure 3.10   The concentration dependence of the chemical shifts of: GpC in H₂O solution (1°C); and ApC in D₂O (4°C, solid circles). The ApC in D₂O chemical shifts are given relative to an external hexamethyldisiloxane capillary reference. Several GpC spectra were also recorded in H₂O solution (4°C, open circles and dotted line) in order to measure the concentration dependence of the amino resonances. (Reproduced with permission from T. Krugh).

Base-paired oligomers of the DNA family

In the last few years, there have been a large number of studies of the double helical properties of many deoxyoligonucleotides. In most of the studies, the shift trends of the exchangeable imino and amino protons and the nonexchangeable base protons have been monitored.

Young and Krugh (1975) demonstrated that the self-complementary deoxy-dimers d-pGpC, d-pCpG, d-GpC and d-CpG and the complementary mixtures d-pGpG + d-pCpC and d-pGpT + d-pApC form Watson-Crick base-paired duplexes

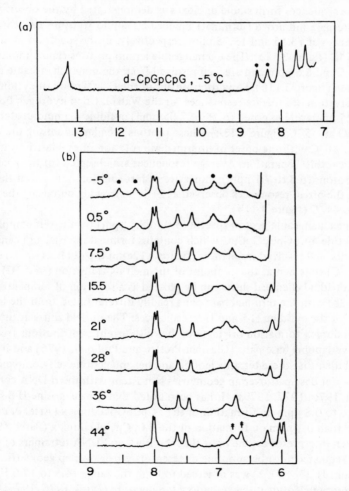

Figure 3.11   The 360 MHz continuous wave proton NMR spectra of 20 mM (in strands) d-CpGpCpG in 0.1 M phosphate, H_2O, neutral pH.[40,50] (a) The spectral region 7 to 14 ppm at $-5\,°C$. The guanosine H-1 Watson-Crick imino proton resonates between 13.0 and 13.5 ppm. (b) The spectral region 6 to 9 ppm as a function of temperature ($-5°$ to $44\,°C$). The cytidine 4-amino protons are designated by * while the guanosine 2-amino protons are designated by †.

at low temperature, by following the downfield shifts of the guanine and cytosine amino resonances as a function of concentration in the range of 1–100 mM. Their results showed the following order in stability: (d-pGpG).(d-pCpC) ⩾ (d-pGpC). d-pGpC) > (d-pCpG).(d-pCpG) > (d-pGpT).(d-pGpT). This reflects the effect of nucleotide sequence and composition on helix stability.

Patel has studied the self-associative properties of sequence isomers d-CpGpCpG (Patel, 1976), d-CpCpGpG (Patel, 1977), d-GpGpCpC (Patel, 1977) and d-GpCpGpC (Patel, 1979a) using ^{1}H and ^{31}P NMR spectroscopy. Due to the two-fold symmetry these tetranucleotides contain equivalent terminal and internal base pairs. These sequences form stable duplexes as demonstrated by the observation of narrow (internal) and broad (terminal) guanosine 1-imino Watson-Crick exchangeable resonances at 13.15 and 13.25 ppm, respectively, in the spectrum of d-CpGp CpG at −5 °C (figure 3.11a). The exchangeable amino protons can be monitored between 6.5 and 8.5 ppm and are well resolved from the nonexchangeable protons in this region (figure 3.11b). Slow rotation about the C—N bond of cytidine results in the observation of separate resonances for the Watson-Crick hydrogen-bonded (8.0 to 8.25 ppm) and exposed (6.75 to 7.05 ppm) cytidine 4-amino protons of d-CpGpCpG at −5 °C (figure 3.11b). These protons broaden on raising the temperature to 21 °C with the onset of intermediate rotation rates relative to the 1.2 ppm chemical shift separation. Average resonances are observed for the guanosine 2-amino protons (6.8 to 7.1 ppm), indicative of fast rotation rates about the C—N bond with the broad resonances observed at 21 °C narrowing on raising the temperature to 44 °C (figure 3.11b).

The nonexchangeable proton spectra (5.0 to 8.5 ppm) for the self-complementary hexanucleotide d-CpGpCpGpCpG which contains terminal internal and central base pairs exhibits well-resolved resonances (figure 3.12a) in the duplex (12 °C) state, the strand (87 °C) state and at the midpoint of the melting transition (67.5 °C) (Patel, 1979a). Individual chemical shifts can be plotted as a function of temperature (figure 3.12b) with the terminal base pairs readily differentiated from the interior base pairs for the cytidine H-5 and H-6 resonances. The upfield shifts at the base protons on duplex formation originate in the ring current contributions from adjacent overlapping base pairs (Giessner-Prettre and Pullman, 1976) and it has been concluded that these tetranucleotide and hexanucleotide self-complementary duplexes adopt base-pair overlap geometries consistent with the B-DNA conformation (Patel, 1976, 1977, 1979a). It has been noted that the guanosine H-8 shifts upfield by 0.6 to 0.8 ppm on formation of the r-CpCpGpG duplex (Arter et $al.$, 1974) but by less than 0.1 ppm on formation of the d-CpCpGpG duplex (Patel, 1977) and this has been interpreted on the basis that the RNA and DNA tetramers of the same base sequence must be adopting different base pair overlap geometries. The coupling sum J1′J2′ + J1′2″ was observed to be in the range 14.5 to 17.0 Hz for the deoxytetranucleotide duplexes at low temperature (Patel, 1976, 1977, 1979a), demonstrating the existence of the ^{2}E sugar pucker as found in B-DNA.

The hexanucleotide d-CpGpCpGpCpG contains five internucleotide phosphates. All five resonances are resolved in the proton noise decoupled phosphorus spectrum in the duplex and strand states between 4.0 and 4.3 ppm upfield from standard tri-

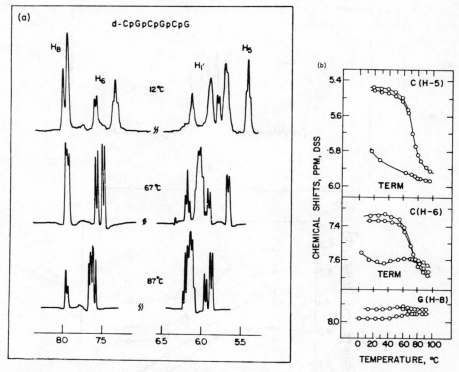

Figure 3.12   (a) The 270 MHz Fourier transform (recorded under conditions of HOD suppression and resolution enhancement) proton NMR spectra of 10.5 mM (in strands) d-CpGpCpGpCpG in 0.1 M cacodylate, 1 mM EDTA, ^2H_2O, pH 6.25 between 5.5 and 8.5 ppm at 12°, 67° and 87 °C. (b) A plot of the base proton chemical shifts between 0° and 100 °C.

methylphosphate (figure 3.13a) (Patel, 1976). The resonances shift downfield on raising the temperature (figure 3.13b) and reflect conformational changes in the sugar phosphate backbone on conversion from stacked duplex to unstacked strands.

Crothers and coworkers have studied the melting transition of the complimentary duplex (d-TpTpGpTpT)·(d-ApApCpApA) by following the shift trends of the methyl resonances (Cross and Crothers, 1974) and the exchangeable imino protons (Crothers, Hilbers and Shulman, 1973) as well as the absorbance melting curve. They observed that the Watson-Crick imino protons broaden out at temperatures below the melting transition of the complex, indicative of exchange with solvent by transient opening of the duplex. The T_m for the pentanucleotide system of Crothers et al. (9 °C) is considerably less than that for the tetramer d-CpCpGpG (42 °C) and this in turn may indicate that the stability of the duplex state depends very much on the number of GC pairs. However, it should be noted that T_m is very sensitive to concentration and ionic strength and a comparison of T_m should be made only under identical conditions.

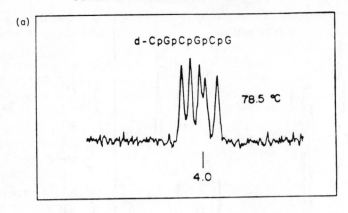

(a)

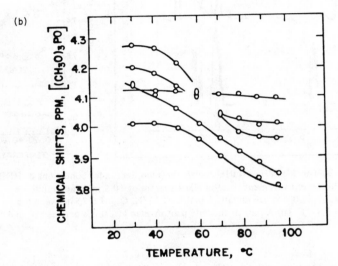

(b)

Figure 3.13   (a) The 145.7 MHz Fourier transform ^{31}P NMR spectrum (proton noise decoupled at 360 MHz) of 10.5 mM (in strands) d-CpGpCpGpCpG in 0.1 M cacodylate, 10 mM EDTA, $^{2}H_2O$, pH 6.5 at 78.5 °C. The chemical shifts are upfield from internal standard trimethyl-phosphate. (b) The temperature dependence of the five internucleotide phosphates are plotted between 30° and 95 °C.

The double helix unwinding mechanism can be probed by studying the transition temperatures at individual base pairs during the melting of an oligonucleotide duplex. For example, in a duplex of the hexamer, there are three kinds of base pairs: terminal, internal and central. The temperature and pH dependence of the resonances from these three kinds of base pairs may reflect the sequence in which the base pairs open.

Patel and coworkers (Hilbers and Patel, 1975; Patel, 1975; Patel and Hilbers, 1975; Patel and Tonelli, 1975) found that the Watson-Crick imino resonances of the self-complementary d-ApTpGpCpApT duplex shifted to higher field and broadened in a sequential order from the ends of the helix on increasing the temperature or raising the pH. The sequential broadening process is consistent with fraying, predominantly at the terminal A·T base pairs and to some extent at the internal A·T base pairs (Patel and Hilbers, 1975). The fraying process represents a rapid opening and closing of the Watson-Crick hydrogen bonds and occurs on a time scale much shorter than the melting transition of the duplex monitored by the central G·C base pairs. The formation of the hexanucleotide duplex was characterised in terms of nucleation at the central G·C base pairs and propagation at the A·T base pairs from additional studies of the exchange process catalysed by phosphate solution (Hilbers and Patel, 1975). NMR studies of the nonexchangeable protons at individual base pairs confirmed the sequential opening of the d-ApTpGpCpApT duplex from its ends (Patel, 1975; Patel and Tonelli, 1975). It is of interest to note that in the duplex of the hexamer d-CpGpCpGpCpG the internal and central base pairs have identical T_m, suggesting that fraying is localised only at the terminal base pair.

Kallenbach and coworkers have studied the Watson-Crick and Hoogsteen hydrogen bonding interactions in double and triple helices in solution (Kallenbach, Daniel and Kaminker, 1976). Their investigation on the self-complementary d-ApApApGpCpTpTpT duplex and the contributions of Patel and Canual (1979) on the self-complementary d-GpGpApApTpTpCpC duplex have extended the NMR investigations to the octanucleotide duplex level. Kearns, Wells and their collaborators have investigated telestability (transmission of stability along the duplex) in block copolymers (Early et al., 1977) and also probed the conformation at the junction of A and B nucleic acid conformations (Selsing et al., 1978). The following reviews are recommended: Kallenbach and Berman (1977) and Kearns (1977) on structural studies of the above oligonucleotide systems; Patel (1979b, c) on NMR studies of synthetic DNAs and synthetic RNAs in solution.

RELATIONSHIP BETWEEN SOLUTION AND SOLID-STATE DATA

Single crystal data are available on several mononucleotides (summarised by Sundaralingam, 1973), three ribodinucleoside monophosphates (Hingerty et al., 1975; Seeman, Day and Rich, 1975; Seeman et al., 1971, 1976; Sussman et al., 1972; Rosenberg et al., 1976), one deoxyribodinucleotide (Camerman, Fawcett and Camerman, 1976), a trimer (Suck et al., 1973), and a tetramer (Viswamitra et al., 1978), and the details of their geometry have been summarised in chapter 2. A comparison between solution and solid-state data is of interest because it provides information about the effect of forces in solution and solid state as well as aggregation on molecular structure.

In the solid state the ribofuranose ring of 3' and 5'-purine and pyrimidine mononucleotides displays either a ^3E or a ^2E pucker with no regularity. The softness of the pucker can be seen from the fact that molecule 1 of 5'ATP is ^3E whereas molecule 2 is ^2E (Kennard et al., 1971; Kim and Sarma, 1978). In aqueous solution they exist as an equilibrium blend of $^2E \rightleftarrows {}^3E$ with the purines showing a slight preference for the ^2E and the pyrimidines for ^3E. For common mononucleotides in the solid state, the magnitude of X_{cn} lies in the *anti* domain, in solution an equilibrium is observed between *anti* and *syn* orientations with preference for *anti* orientation. The C3'—O3' torsion of ribo-3'-mononucleotides lies in a broad range from 237–269° in the solid state. In solution the C3'—O3' torsion is coupled to sugar ring conformation by the equilibrium $^2E\alpha^+ \rightleftarrows {}^3E\alpha^-$ where α^+ and α^- are centred around 275 and 205°.

In addition, the overlap features of adjacent molecules in crystals and aqueous solution show conspicuous differences. Evans and Sarma (1974) have shown that the specific intermolecular base stacking geometries in 5'-AMP solutions do not agree with those reported in 5'-AMP crystals (Bugg, 1971; Kraut and Jensen, 1963). It was also shown that the base-stacking orientations of neutral adenosine in the solid state are considerably different from the average orientations in aqueous solution. One reason for this is that in aqueous solution bases aggregate due to base-stacking interactions and not due to hydrogen bonding. In the adenosine crystal (Lai and Marsh, 1972) there is a myriad of interstack, intermolecular hydrogen bonds; for example, each nucleoside is hydrogen bonded to six neighbouring molecules. This strong interaction, which is peculiar to the solid state, is expected to stabilise otherwise unnatural intermolecular orientations. Furthermore, it is expected that such hydrogen bonding in some cases may alter the preferred intramolecular conformation as well. For example, in the case of adenosine there is a hydrogen bond stabilising the $\epsilon \simeq 180°$ orientation about the C4'—C5' bond, whereas in aqueous solution it is clear that the $\epsilon \simeq 60°$ orientation is strongly preferred. The differences in the immediate environments most likely account for the observed drastic discrepancy between the solid-state and solution conformations of 6-thiopurine riboside and 4-thiouridine (Evans and Sarma, 1975; Saenger and Scheit, 1970; Shefter, 1968).

Despite these differences between solid-state and solution conformations of mononucleotides, there is an area of excellent agreement. In both solution and solid-state studies it is invariably found that the C4'—C5' and C5'—O5' bonds form a stable conformational network in which the magnitudes of δ and ϵ are, respectively, $\simeq 60°$ and $\simeq 180°$. This suggests the possibility that in solution oligonucleotides may achieve conformational versatility by torsional variation about X_{cn}, C3'—O3', O3'—P, O5'—P bonds as well as by variation in ribose pucker.

Earlier it was noted that the first step in oligomerisation, that is, monomer to dimer, causes major changes in conformation and, hence, the dimers attempt to become conformationally pure. Because of this we may expect a great deal of common stereochemical features between the solution and solid state conformations for the higher oligomers. In fact the preferred spatial configuration for GpC, ApU and d-pTpT in solution is the same as that reported for the corresponding molecules in the solid state (chapter 2). Further, the solid state data on UpA and ApA$^+$pA$^+$, where g^+g^+ and tg^- phosphodiester torsions have been detected, can be considered to be in agreement with the solution findings that in addition to the g^-g^- helical arrays, in which the phosphodiester assumes different arrangements such as g^+g^+, tg^-, g^+ are allowed. However, the solution data on d-ApT and d-TpA cannot correctly predict the observed solid-state structure for d-pApTpApT which is characterised by alternating sugar puckers as well as g^-g^-, tg^-, and g^-g^- phosphodiester torsions. To date, no solution data of this tetramer are available.

ACKNOWLEDGMENTS

This research was supported by grant CA12462 from National Cancer Institute of NIH, grant PCM-7822531 from National Science Foundation, and grant 1-PO7-PR-PR00798 from the Division of Research Resources. The authors thank Thomas R. Krugh for providing figure 3.10 and Dinshaw J. Patel for providing figures 3.11, 3.12 and 3.13. We are grateful to Dinshaw J. Patel for his extensive revision of the section on base-paired oligonucleotides of the DNA family.

REFERENCES

Arter, D. B., Walker, G. C., Uhlenbeck, O. C. and Schmidt, P. G. (1974). *Biochem. biophys. Res. Commun.*, **61**, 1089.

Bangerter, B. W. and Chan, S. I. (1969). *J. Am. chem. Soc.*, **91**, 3910.

Berman, H. M. and Neidle, S. (1979). In *Stereodynamics of Molecular Systems* (ed. R. H. Sarma), Pergamon Press, Oxford, p. 367.

Borer, P. N., Kan, L. S. and Ts'o, P. O. P. (1975). *Biochemistry*, **14**, 4847.

Bugg, C. E. (1971). In *Purines: Theory and Experiment* (ed. B. Pullman), Academic Press, New York.

Camerman, N., Fawcett, J. K. and Camerman, A. (1976). *J. molec. Biol.*, **107**, 601.

Chan, S. I. and Nelson, J. H. (1969). *J. Am. chem. Soc.*, **91**, 168.

Cheng, D. M., Dhringra, M. M. and Sarma, R. H. (1978). *Nucl. Acids Res.*, **5**, 4399.

Cheng, D. M. and Sarma, R. H. (1977a). *J. Am. chem. Soc.*, **99**, 7333.

Cheng, D. M. and Sarma, R. H. (1977b). *Biopolymers*, **16**, 1687.

Cross, A. D. and Crothers, D. M. (1974). *Biochemistry*, **10**, 6015.

Crothers, D. M., Hilbers, C. W. and Shulman, R. G. (1973). *Proc. natn. Acad. Sci. U.S.A.*, **70**, 2899.

Danyluk, S. S., Ainsworth, C. F. and MacCross, M. (1978). In *Nuclear Magnetic Resonance in Molecular Biology* (ed. B. Pullman), Reidel, Holland, p. 111.

Dhingra, M. M. and Sarma, R. H. (1979a). In *Stereodynamics of Molecular Systems* (ed. R. H. Sarma), Pergamon Press, Oxford, p. 3.

Dhingra, M. M. and Sarma, R. H. (1979b). *Int. J. Quant. Chem. Quantum Biol. Symp.* (in press).

Dhingra, M. M., Sarma, R. H., Geissner-Prettre, C. and Pullman, B. (1978). *Biochemistry*, **17**, 5815.

Early, T. A., Kearns, D. R., Burd, J. F., Larson, J. E. and Wells, R. D. (1977). *Biochemistry*, **16**, 541.

Evans, F. E., Lee, C. H. and Sarma, R. H. (1975). *Biochem. biophys. Res. Commun.*, **63**, 106.

Evans, F. E. and Sarma, R. H. (1974). *Biopolymers*, **13**, 2117.

Evans, F. E. and Sarma, R. H. (1975). *J. Am. chem. Soc.*, **97**, 3215.

Evans, F. E. and Sarma, R. H. (1976). *Nature, Lond.*, **263**, 567.

Ezra, F. W., Lee, C. H., Kondo, N. S., Danyluk, S. S. and Sarma, R. H. (1977). *Biochemistry*, **16**, 1977.

Giessner-Prettre, C. and Pullman, B. (1976). *Biochem. biophys. Res. Commun.*, **70**, 578.

Hilbers, C. W. and Patel, D. J. (1975). *Biochemistry*, **14**, 2656.
Hingerty, B., Sundramanian, E., Stellman, S. D., Broyde, S. B., Sato, T. and Langridge, R. (1975). *Biopolymers*, **14**, 227.
Hruska, F. E. and Danyluk, S. S. (1968). *J. Am. chem. Soc.*, **91**, 168.
Kallenbach, N. R. and Berman, H. M. (1977). *Q. Rev. Biophys.*, **6**, 477.
Kallenbach, N. R., Daniel, W. E. Jr., and Kaminker, M. A. (1976). *Biochemistry*, **15**, 1218.
Kan, L. S., Barrett, J. C. and Ts'o, P. O. P. (1973). *Biopolymers*, **12**, 2409.
Kan, L. S., Borer, P. N. and Ts'o, P. O. P. (1975). *Biochemistry*, **14**, 4865.
Kan, L. S., Ts'o, P. O. P., Van der Haar, F., Sprinzel, M. and Cramer, F. (1975). *Biochemistry*, **14**, 3278.
Kearns, D. R. (1977). *A. Rev. Biophys. Bioeng.*, **6**, 477.
Kennard, O., Isaacs, N. W., Motherwell, W. D. S., Coppola, J. C., Wampler, D. L., Larson, A. C. and Watson, D. G. (1971). *Proc. R. Soc. A.*, **325**, 401.
Kim, C. H. and Sarma, R. H. (1978). *J. Am. chem. Soc.*, **100**, 1571.
Kraut, J. and Jensen, L. H. (1963). *Acta crystallogr.*, **16**, 79.
Krugh, T. R., Laing, J. W. and Young, M. A. (1976). *Biochemistry*, **15**, 1224.
Ladner, J. E., Jack, A., Robertus, J. D., Brown, R. S., Rhodes, D., Clark, B. F. C. and Klug, A. (1975). *Nucl. Acids Res.*, **2**, 1623.
Lai, T. F. and Marsh, R. E. (1972). *Acta crystallogr.*, **B28**, 1982.
Lee, C. H., Evans, F. E. and Sarma, R. H. (1975). *FEBS Lett.*, **51**, 73.
Lee, C. H., Ezra, F. S., Kondo, N. S., Danyluk, S. S. and Sarma, R. H. (1977). *Biochemistry*, **16**, 1977.
Lee, C. H., Ezra, F. S., Kondo, N. S., Sarma, R. H. and Danyluk, S. S. (1976). *Biochemistry*, **15**, 3627.
Neidle, S., Taylor, G., Sanderson, M., Shieh, H. S. and Berman, H. M. (1978). *Nucl. Acids Res.*, **5**, 4417.
Olson, W. K. (1977). *Proc. natn. Acad. Sci. U.S.A.*, **74**, 1775.
Patel, D. J. (1975). *Biochemistry*, **14**, 3985.
Patel, D. J. (1976). *Biopolymers*, **15**, 533.
Patel, D. J. (1977). *Biopolymers*, **16**, 1635.
Patel, D. J. (1979a). *Biopolymers*, **18**, 553.
Patel, D. J. (1979b). *Acct. Chem. Res.*, **12**, 118.
Patel, D. J. (1979c). In *Nucleic Acid Geometry and Dynamics* (ed. R. H. Sarma), Pergamon Press, Oxford (in press).
Patel, D. J. and Canuel, L. L. (1979). *Eur. J. Biochem.*, **96**, 267.
Patel, D. J. and Hilbers, C. W. (1975). *Biochemistry*, **14**, 2651.
Patel, D. J. and Tonelli, E. A. (1975). *Biochemistry*, **14**, 3991.
Quigley, G. J., Seeman, N. C., Wang, A. H. J., Suddath, F. L. and Rich, A. (1975). *Nucl. Acids Res.*, **2**, 2329.
Rich, A., Quigley, G. J. and Wang, A. H. J. (1979). In *Stereodynamics of Molecular Systems* (ed. R. H. Sarma), Pergamon Press, Oxford, p. 315.
Rosenberg, J. M., Seeman, N. C., Day, R. O. and Rich, A. (1976). *J. molec. Biol.*, **104**, 145.
Saenger, W. (1979) and Scheit, K. H. (1970). *J. molec. Biol.*, **50**, 153.
Sarma, R. H. (1979). In *Nucleic Acid Geometry and Dynamics* (ed. R. H. Sarma), Pergamon Press, Oxford, p. 3.
Sarma, R. H. and Danyluk, S. S. (1977). *Int. J. Quantum Chem. Quantum Biol. Symp.*, **4**, 269.
Sarma, R. H., Dhingra, M. M., Cheng, D. M., Ezra, F. S., MacCross, M. and Danyluk, S. S. (1980). *Biochemistry*, **19**, 2491.
Sarma, R. H., Dhingra, M. M. and Feldman, R. J. (1979a). In *Nucleic Acid Geometry and Dynamics* (ed. R. H. Sarma), Pergamon Press, Oxford, p. 83.

Sarma, R. H., Dhingra, M. M. and Feldman, R. J. (1979*b*). In *Stereodynamics of Molecular Systems* (ed. R. H. Sarma), Pergamon Press, Oxford, p. 153.

Seeman, N. C., Day, R. O. and Rich, A. (1975). *Nature, Lond., 253*, 324.

Seeman, N. C., Rosenberg, J. M., Suddath, F. L., Kim, J. P. and Rich, A. (1976). *J. molec. Biol., 104*, 109.

Seeman, N. C., Sussman, J. L., Berman, H. M. and Kim, S. H. (1971). *Nature, Lond., 233*, 90.

Selsing, E., Wells, R. D., Early, T. A. and Kearns, D. R. (1978). *Nature, Lond., 275*, 249.

Shefter, E. (1968). *J. Pharm. Sci., 57*, 1157.

Singh, H., Herbut, M. H., Lee, S. H. and Sarma, R. H. (1976). *Biopolymers, 15*, 2167.

Stout, C. D., Mizuno, H., Rubin, J., Brennan, T., Rao, S. T. and Sundaralingam, M. (1976). *Nucl. Acids Res., 3*, 1111.

Suck, D., Manor, P. C., German, G., Schwalbe, C. H., Weimann, G. and Saenger, W. (1973). *Nature, Lond., 246*, 161.

Sundaralingam, M. (1973). *Jerusalem Symp. Quantum Chem. Biochem., 5*, 417.

Sundaralingam, M. (1975). In *Structure and Conformation of Nucleic Acid and Protein-Nucleic Acid Interactions* (ed. M. Sundaralingam and S. T. Rao), University Park Press, Baltimore, p. 487.

Sussman, J. L. and Kim, S. H. (1976). *Biochem. biophys. Res. Commun., 68*, 89.

Sussman, J. L., Seeman, N. C., Kim, S. H. and Berman, H. M. (1972). *J. molec. Biol., 66*, 403.

Ts'o, P. O. P., Kondo, N. S., Schweizer, M. P. and Hollis, D. P. (1969). *Biochemistry, 8*, 997.

Viswamitra, M. A., Kennard, O., Jones, P. G., Sheldrick, G. M., Salisbury, S., Falvello, L. and Shakked,·Z. (1978). *Nature, Lond., 273*, 687.

Young, M. A. and Krugh, T. R. (1975). *Biochemistry, 14*, 4841.

4

The secondary structures of polynucleotide chains as revealed by x-ray diffraction analysis of fibers

Struther Arnott

INTRODUCTION

To date, details of the helical conformations of nucleic acids and analogous synthetic polynucleotides have been derived mainly from x-ray diffraction analyses of fibrous specimens. The diffraction from oriented (and sometimes polycrystalline) fibres of the polynucleotides is insufficient by itself to define conformations very accurately since diffraction signals corresponding to spacings less than 3 Å are scarce. However, it is possible to supplement the diffraction data from the polymers with stereo-chemical information in the form of bond lengths, bond angles, hydrogen bond distances, and base and furanose ring shapes obtained from very accurate x-ray diffraction analyses of nucleotides and other relevant small molecules. When this is done the relative atomic positions in the polymers can be determined with an accuracy of a few hundredths of an angstrom, the limitation of accuracy coming largely from the uncertainties in the fixed values of whatever stereochemical features have been imposed on the polymer models. Before 1966 the stereochemical constraints were imposed by building scale wire models (for example, Crick and Watson, 1954; Langridge *et al.*, 1960). After the introduction of the linked-atom least-squares procedures by Arnott and Wonacott (1966) computer methods have improved both the precision and the accuracy of fibre diffraction analyses (Smith and Arnott, 1978).

The importance of assigning correct, fixed, stereochemical features is amply illustrated by the history of models for the most common conformation of the DNA duplex, *B*-DNA. All models for this structure have been constrained to have the tenfold (10_1) right-handed screw symmetry and 34 Å pitch indicated by the fibre diffraction, and to have the hydrogen-bonded complementary base pairing proposed by Watson and Crick (1953). However, the first model (Crick and Watson,

65

Table 4.1 Conformation angles in successive models for *B*-DNA (the classical DNA duplex): (1) Crick and Watson (1954), (2) Langridge *et al.* (1960), (3) Arnott and Hukins (1972), (4) Arnott and Chandrasekaran (1980*a*)

Conformation angle	(1)	(2)	(3)	(4)
α (C4′—C3′—O3′—P)	−174	147	−166	−133
β (C3′—O3′—P—O5′)	−64	−78	−136	−157
γ (O3′—P—O5′—C5′)	−178	−79	−25	−41
δ (P—O5′—C5′—C4′)	179	−148	160	136
ϵ (O5′—C5′—C4′—C3′)	155	58	27	38
ζ (C5′—C4′—C3′—O3′)	100[a]	130[b]	146[c]	139[d]
χ (O4′—C1′—N9—C4)	−171	−93	−98	−102

[a]Corresponds to a fixed C3′-*endo* puckered furanose ring.

[b]Associated with a not very probable C3′-*exo* or C2′-*endo* puckered furanose.

[c]Corresponds to fixed standard (that is, most probable) C2′-*endo* puckered furanose.

[d]The sugar ring was allowed to vary its shape within the ranges observed with mononucleotide structures.

1954) had sugar rings puckered C3′-*endo* as in Furberg's (1950) structure of cytidine, which was then the only nucleoside whose structure had been determined crystallographically. In consequence, the base pairs had to be placed about 5 Å from the helix axis. As Wilkins and coworkers found (Langridge *et al.*, 1957, 1960), such a model gives a very poor fit with the observed intensities of diffraction and by 1960 they had produced a refined model with base pairs much nearer the helix axis (as shown in figure 4.1b) which required sugar rings puckered rather differently as well as changes in the conformations at the P—O5′ and C5′—C4′ bonds (table 4.1). In the more refined models of *B*-DNA (Arnott and Hukins, 1972*a*, 1973; Arnott and Chandrasekaran, 1980*a, b*) the sugar rings have more standard (Arnott and Hukins, 1972*b*) C2′-*endo* shapes and more probable values for bond lengths and bond angles than in Wilkins' models, but the general disposition of the main scattering groups is essentially the same.

The development of procedures for producing optimised molecular and crystal models with explicitly defined characteristics has not been of merely operational significance. It is not usually possible to use methods such as isomorphous replacement objectively to solve the x-ray phase problem in fibre diffraction analyses. Instead approximate phases are implicitly provided by postulating some initial structural model and improved by the refinement of this model. However, there is nothing in such a process which would automatically remedy an initial choice of the wrong kind of model. Where another model can be envisioned there is no remedy for the dilemma except to produce the optimum model of each kind then to compare the x-ray diffraction and stereochemical properties of each optimised model to determine whether one is clearly preferable. The importance of being able to do this conveniently for any fibrous system cannot be too strongly emphasised, especially when we consider that in the long (30-year) history of nucleic acid structure analyses no structures, except those produced by Maurice Wilkins' school, have survived reappraisal. As well as the fact that the original Crick and Watson (1954) model for *B*-DNA contained C3'-*endo* puckered furanose rings which are now associated with the *A* family of polynucleotide structures it is also known that:

(1) The structure of reovirus RNA duplexes is not a tenfold double helix as suggested by Langridge and Gomatos (1963) but an elevenfold double helix (Arnott *et al.*, 1967, 1973).

(2) The structure of the eightfold helical duplex of poly d(IC)·poly d(IC) is not the bizarre left-handed model proposed by Mitsui *et al.* (1970) but a right-handed, conformationally unexceptional member of the *B* family of polynucleotides (Arnott *et al.*, 1974).

(3) The multistranded complex formed with poly(I) or poly(G) is not triple-stranded as suggested by Rich (1958) but quadruple-stranded (Arnott, Chandrasekaran and Marttila, 1974; Zimmerman *et al.*, 1975; Chou *et al.*, 1977).

(4) The poly(C) helices trapped in oriented, polycrystalline fibres by Langridge and Rich (1963) are not duplexes with coaxial, base-paired, twelvefold helical poly-nucleotide chains but single, sixfold helices with no direct base–base interactions (Arnott *et al.*, 1976).

It has been suggested that physical probes other than x-ray analyses of fibres might provide more accurate details of DNA structure. Ultraviolet circular dichroism spectra are indeed very sensitive to the secondary structures present but they are also sensitive to the base sequences present in a way that currently is being disputed (Gray and Tinoco, 1970; Arnott, 1975). Certainly, it was an unwise interpretation of a circular dichroism spectrum which led directly to a structure of the wrong chirality being proposed for poly d(IC)·poly d(IC) (Mitsui *et al.*, 1970). It is more likely, therefore, that this technique will prove more useful in deciding which (previously defined) structure is present and will be a complementary probe to x-ray diffraction for studies of DNA in solution. Infrared linear dichroism has been used to define the orientation of P––O and other bonds in DNA (Pilet and Brahms, 1972; Pilet *et al.*, 1973). Nominally this technique should provide more accurate results than the x-ray fibre diffraction analyses. However, it appears that the conclusions from the infrared studies are not generally compatible with the x-ray results. Moreover, there is every reason to doubt some of the assumptions that are necessary for the infrared calculations (Beetz *et al.*, 1979). As with ultraviolet dichroism, it is more likely that infrared linear dichroism will prove more useful in monitoring the occurrence of structural transitions rather than in defining accurately features of the molecular structure being examined. NMR studies may turn out to be the most useful ancillary to x-ray work. The qualitative conformational predictions made by Patel *et al.* (1979) from their NMR study of the salt-induced transition in oligo d(GC)·oligo d(GC) were amply confirmed by the crystal structure of d(CG)$_3$ (Wang *et al.*, 1979) and the fibre structure of poly d(GC)·poly d(GC) (Arnott *et al.*, 1980*a*).

DESCRIBING POLYNUCLEOTIDE HELICAL STRUCTURES

The backbone conformation angles ($\alpha, \ldots, \zeta$) and the glycosidic torsion angle (χ) are useful in defining the degree of relatedness between the different polynucleotide secondary structures. The bond torsions of the sugar rings and their endocyclic bond angles are nowadays all treated as stiffly elastic variables but for purposes of classification they are represented sufficiently by ζ.

The backbone conformation angles can be described as *gauche* (g^- or g^+) or *trans* (t), and the glycosidic torsion as *anti* (a) or *syn* (s), as in earlier chapters. Consequently, the nucleotide shapes of, say, B-DNA and its close relatives can be described by a seven-letter conformational code ttg^-tg^+ta designating the successive conformations of ($\alpha, \ldots, \zeta, \chi$). This leads to a convenient taxonomic classification by which the helical secondary structures of polynucleotides are divided first into families, depending on their sugar ring puckering (since this has the biggest impact on gross morphology), then into genera in which all the molecular species have the same *kind* of conformation angle at each bond. For DNA and RNA complexes with Watson-Crick base pairing the known families of secondary structures and some of their characteristics are shown in tables 4.2a and 4.2b.

Table 4.2a Classification of double helical poly(mono)nucleotide duplex structures determined by x-ray diffraction analysis of fibres

Family	A		B	
Furanose conformation	C3′-endo		C2′-endo	
Conformational genera	$tg^-g^-tg^+g^+a$	tg^-tttg^+a	ttg^-tg^+ta	$tttttta$
Number of congeneric species	16	1	4	1
Helical characteristics h (Å)	2.6–3.3	3.1	3.0–3.4	3.3
t (°)	30.0–32.7	36.0	36.0–45.0	48.0

Table 4.2b Classification of double helical poly(di)nucleotide structures determined by x-ray fibre diffraction analysis

Family	A + B		
Furanose conformation	C3′-endo + C2′-endo		
Conformational genera	$g^-g^+g^+ttg^+s$ $+ g^-g^-g^-tg^+ta$	$g^-g^+g^+ttg^+s$ $+ tg^+g^-tg^-ta$	$g^-g^+g^+ttg^+s$ $+ ttg^-ttta$
Number of congeneric species	2	1	1
Helical characteristics h (Å)	7.3–7.6	7.2	10.2
t (°)	(−60)–(−51.4)	−25.7	0

Molecular helices can be described also by their pitch, screw symmetry and the number of residues per pitch. In this context a residue is the asymmetric unit of structure and may be nucleotide, dinucleotide, and so on. When comparing structures it is convenient to know the axial translation per residue (h), which is the ratio of the pitch (P) and the number of residues per pitch, and also the rotation per residue (t), which is (h/P) 360°. Many morphological characteristics are strongly correlated with the values of h and t, including the tilt of the bases and the widths and depths of the major and minor grooves in duplexes.

MORPHOLOGY OF POLYNUCLEOTIDE DUPLEXES WITH WATSON-CRICK BASE PAIRING

Figure 4.1 shows the two extreme variants of the A-genus with conformations $tg^-g^- tg^+g^+a$. A-DNA with (h, t) = (2.6 Å, 32.7°) was one of first two polymorphs of DNA to be described (Franklin and Gosling, 1953). A''-DNA has (h, t) = (3.3 Å, 30°) and was observed in the Watson-Crick base-paired portion of the triplex

Table 4.3 Grooves in polymononucleotide duplex helices

Family	t (°)	h (Å)	m (Å)	M (Å)	d (Å)	D (Å)
A	30.0	3.26	10.9	10.0	4.2	12.6
	32.7	2.56	11.0	2.7	2.8	13.5
B	36.0	3.37	5.7	11.7	7.5	8.5
	45.0	3.03	1.3	8.9	6.7	5.8

t = rotation per nucleotide, h = axial translation per nucleotide, m and M = widths of minor and major grooves, d and D = depths of minor and major grooves.
The groove widths are defined as the perpendicular separations of the two helices which connect the phosphorus atoms of each strand, reduced by 5.8 Å (the diameter assumed for a phosphate group).
Groove depths are defined as follows: $d = r_p - r_{N2G} + 1.4$ for both the A and B genera; $D = r_p - r_{N6A} + 1.4$ for the B genus; $D = r_p + r_{N6A} + 1.4$ for the A genus. r_p, r_{N2G} and r_{N6A} are the cylindrical polar radii of the phosphorus, guanine N2 and adenine N6 atoms respectively. The constant term (1.4 Å) represents the difference of the phosphate radius (2.9 Å) and the van der Waals radius of nitrogen (1.5 Å).

poly dT·poly dA·poly dT (Arnott and Selsing, 1974). In both molecular species the minor groove (in which the glycosidic links of a base pair make an obtuse angle with one another) is shallow because the base pairs are positioned 'forward'. The major groove (where the 'back' of the base pairs is visible) is correspondingly deep. The marked difference in the value of h of the two congeners is correlated with the tilt of the base pairs which make an angle ~70° to the helix axis in A-DNA and ~85° in A''-DNA. Related to this is the width of the major groove (table 4.3) which increases markedly with h.

These two structures demonstrate quite vividly the marked morphological

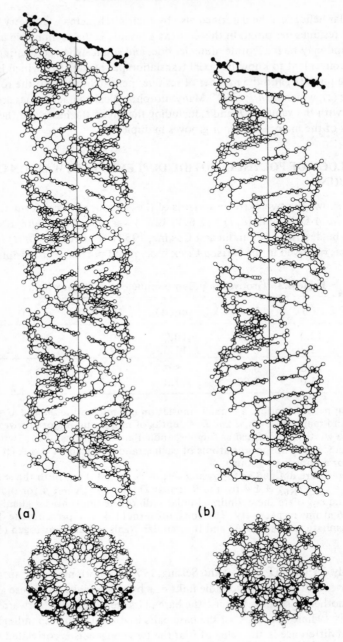

Figure 4.1 Mutually perpendicular projections of model fragments of (a) A-DNA and (b) A''-DNA. Both molecular species belong to the same genus of the A family with nucleotide conformations $tg^- g^- tg^+ g^+ a$ but possess extreme values of the axial translation per nucleotide residue: h = 2.6 Å in the former, h = 3.3 Å in the latter. Rotations per residue are similar: 32.7° in A-DNA and 30.0° in A''-DNA.

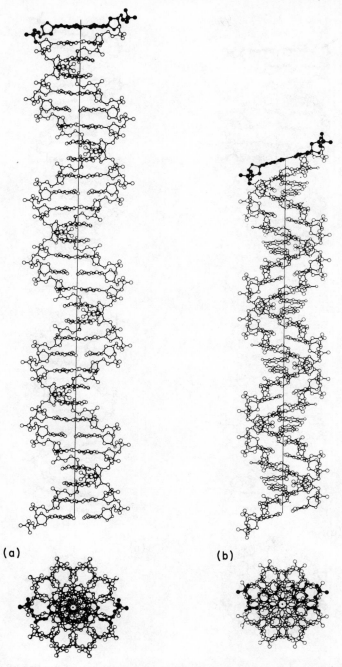

Figure 4.2 Projections of (a) *B*-DNA and (b) *D*-DNA structures which represent the extremes of morphological variation within the *B* family genus with nucleotide conformations of the type *ttg⁻ tg⁺ ta*. The values of (*h*, *t*) are (3.4 Å, 36.0°) for *B*-DNA and (3.0 Å, 45.0°) for *D*-DNA.

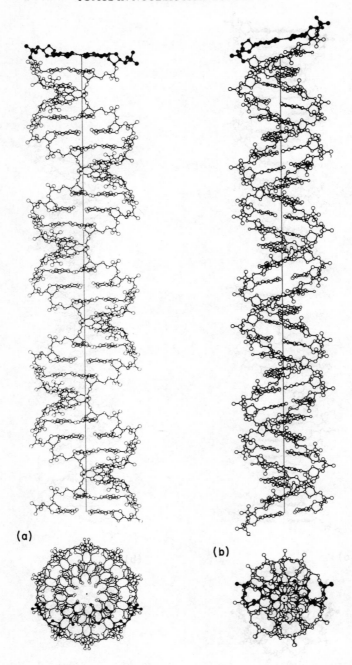

Figure 4.3 Projections of the sole representatives of the rarer genera of the A and B families. In (a) the structure has nucleotide conformations (A) $tg^-\ tttg^+a$ and $(h,\ t) = (3.1$ Å, $36.0°)$; in (b) the conformations are (B) $tttttta$ and $(h,\ t) = (3.3$ Å, $48.0°)$.

differences which can be contrived without any major conformational transition occurring at any bond.

In the genus just discussed the rotations per residue are confined to a narrow range: $30.0° \leqslant t \leqslant 32.7°$. This restriction vanishes for single-stranded structures like poly C (Arnott *et al.*, 1976) which has $t = 60°$ although conformationally it belongs to the same genus. For greater rotations per residue in duplexes of the A family, conformational transitions to the tg^-ttg^+a genus are necessary. At this time this genus is represented by the single secondary structure (figure 4.3a) observed with poly d(I)·poly r(C) (Banerjee *et al.*, 1980). This 10_1 helix with $h = 3.1$ Å has the curious distinction of being very similar to the specific structure Crick and Watson (1954) published for B-DNA. Comparison of this A family structure (figure 4.3a) with the contemporary model of B-DNA (figure 4.2a) shows that there are considerable similarities in the orientations of the base pairs and the widths of the major and minor grooves. However, since the base pairs in all members of the A-family are distant from the helix axis the minor groove is much shallower and the major groove correspondingly deeper than in B-DNA ($\Delta d \sim \Delta D \sim 4$ Å).

B-DNA is the eponymous member of the B family and belongs to the conformational genus ttg^-tg^+ta. In this family the change to C2'-*endo* furanose rings produces morphological variation quite different from the A family. B-DNA itself, with the largest rise and lowest rotation per residue in this genus, has its base pairs astride the helix axis and almost perpendicular to it. The major and minor grooves are thus about equally deep. Both are also wide (table 4.3). As h decreases and t increases in other species of this genus, the minor groove becomes deeper and narrower and the bases tilt in the opposite direction from A-DNA. At the opposite extreme from B-DNA we have D-DNA (Arnott *et al.*, 1976) (figure 4.2b) with $(h, t) = (3.0$ Å$, 45.0°)$. It has a notably compact structure: not only is the mass per unit length 10 per cent greater than in B-DNA, the diameter is markedly less.

Rotations per residue greater than $45°$ require conformational transitions to the new genus $ttttg^+a$ so far represented by only one species, observed with poly d(IIT)·poly d(ACC) and poly d(AIT)·poly d(ACT) (Leslie *et al.*, 1980). This structure with the trivial name E-DNA has $(h, t) = (3.3$ Å$, 48.0°)$ and, therefore, the largest (positive) rotation per residue observed in any DNA duplex.

Only recently duplexes with markedly reduced (zero and negative) rotations (Arnott *et al.*, 1980a; Arnott, Bond and Chandrasekaran, 1980; Arnott and Chandrasekaran, 1980b) have been discovered. All belong to a new family with a dinucleotide as the repeating structural motif. The two nucleotides comprising this have vert different conformations including differently puckered sugar rings. Thus we can call this family $(A+B)$ (table 4.2b).

The left-handed helix observed with poly d(GC)·poly d(GC) and poly d(AC)· poly d(GT) (Arnott *et al.*, 1980a) belongs to the conformational genus $g^-g^+g^+ttg^+s$ $+g^-g^-g^-tg^+ta$, indicating an asymmetric unit where one nucleotide has familiar conformations but the other has the unusual g^+g^+ conformation for the phosphate diester and the rare *syn* nucleoside conformation. What appears to be a variant of this S-DNA conformation is observed with poly d(As^{4}T)·poly d(As^{4}T) where h has increased from 7.3 Å to 7.6 Å and t from $(-60.0°)$ to $(-51.4°)$.

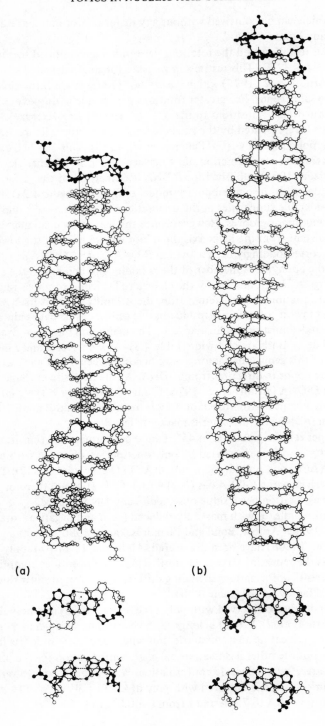

(a)   (b)

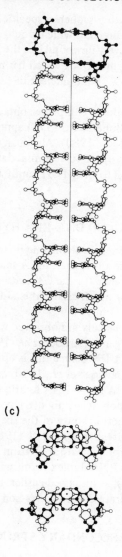

(c)

Figure 4.4 Projections of representatives of genera of the ($A + B$) family. In this family the repeated motif is a dinucleotide in which alternate nucleotides have rather different conformations including (but not limited to) those of the sugar rings. The views parallel to the helix axes show the notably different overlap in successive dinucleoside phosphates. In all three structures an alternating purine/pyrimidine base sequence is implied but this is required only in the structure in (a). The conformations and (h, t) values are: (a) $g^- g^+ g^+ t t g^+ s + g^- g^- g^- t g^+ a$ and (7.3 Å, $-60.0°$); (b) $g^- g^+ g^+ t t g^+ s + t g^+ g^- t g^- t a$ and (7.2 Å, $-25.7°$); (c) $g^- g^+ g^+ t t g^+ s + t t g^- t t t a$ and (10.2 Å, $0.0°$).

S-DNA is unusual in being rather sequence specific, namely alternating purine/pyrimidine. Left-handed DNA duplexes which would accommodate any base sequence would have to be relatively unwound like the $g^-g^+g^+ttg^+s + tg^+g^-tg^-ta$ structure observed with calf thymus DNA which has been stretched and allowed to relax partly. Here t is only $-25.7°$ per dinucleotide (Arnott and Chandrasekaran, 1980b) (table 4.2b).

The DNA duplexes in the ordered complexes obtained with some platinum metallointercalation reagents (Bond $et\ al.$, 1975) appear to be completely unwound (Arnott, Bond and Chandrasekaran, 1980). Although completely base-paired, every second stack has been disrupted to accommodate the intercalating agent. Again the conformation belongs to the $(A+B)$ family but a separate genus $g^-g^+g^+ttg^+s + ttg^-ttta$.

STRUCTURES OF RNA–RNA AND DNA–RNA HYBRID DUPLEXES

Watson-Crick base-paired duplexes involving one or two RNA chains have never been observed with structures in other than the A family. This may be because ribose rings have a slight preference for the C3′-$endo$-puckered form or (more likely) because the C2′ position which carries the additional $-$OH group is sterically restricted in duplexes of the B family.

In fibres the RNA duplexes usually are observed with structures near the middle of the range of the $tg^-g^-tg^+a$ genus (Arnott $et\ al.$, 1967, 1968), but RNA chains can also be isomorphous with A-DNA since this is observed with DNA–RNA hybrids (Milman $et\ al.$, 1966; Banerjee $et\ al.$, 1980) and in helical regions of transfer RNA (Ladner $et\ al.$, 1974). A DNA–RNA hybrid like A'-RNA with $h = 3.0$ Å and $t = 30°$ was described for poly r(I)·poly d(C) by O'Brien and McEwan (1970). In a more extensive fibre diffraction study of the six homopolymer hybrid duplexes involving A:U, G:C and I:C pairs, Banerjee $et\ al.$ (1980) showed that in general RNA–DNA hybrids are isomorphous with A-DNA in conditions of low relative humidity but mimic the RNA–RNA duplex forms with $h \sim 3$ Å at high relative humidities. Poly d(I)·poly r(C), as discussed earlier, is unusual in having a rare structure much like the inaccurate Crick and Watson (1954) model for B-DNA.

OTHER POLYNUCLEOTIDE SECONDARY STRUCTURES

Poly(I) and poly(G) form isomorphous, quadruple-stranded complexes (Arnott, Chandrasekaran and Marttila, 1974; Zimmerman $et\ al.$, 1975) and not the triple-stranded structure envisaged earlier by Rich (1958). The four parallel polynucleotide chains are identical and belong to the $tg^-g^-tg^+g^+a$ genus of conformations. Indeed all the multistranded polynucleotide systems have helical chains of this type whether all the chains are RNA as in poly U·poly A·poly U (Arnott and Bond, 1973a) or DNA as in poly dT·poly dA·poly dT (Arnott and Selsing, 1974). Neither the presence of Hoogsteen type A:U(T) pairs where the interglycosidic link distance C1′ . . . C1′ is 2 Å shorter than for a Watson-Crick pair, nor the presence of I:A pairs with C1′ . . . C1′ 2 Å longer, preclude polynucleotide chains from having this common A conformation. The structures of poly I·poly A·poly I (Arnott and

Bond, 1973*b*) and poly X·poly X (Arnott *et al.*, 1980*b*) are striking examples of this. One form of the poly X·poly X duplex and also the poly (s^{2}U)·poly (s^{2}U) duplex (Arnott *et al.*, 1980*c*) are even more remarkable in contriving to accommodate in one case Pu:Pu and in the other Py:Py base pairs in double helices where the chains are *isomorphous* with those of *A*-DNA (although, of course, their mutual arrangement is different).

It has long been recognised that the hydrogen bonds are not the main factors stabilising polynucleotide secondary structures: favourable conformations at each single bond and (hydrophobic) base-stacking interactions are much more important. It would not be surprising, therefore, to find stable polynucleotide structures with no hydrogen-bonded base–base interactions. So far, however, there are only two examples of such structures. The rhombohedral form of poly(C) contains sixfold poly(C) helices $(h, t) = (3.1 \text{ Å}, 60.0°)$. The only interactions between the non-coaxial chains involve the free —OH groups of the ribose rings (Arnott *et al.*, 1976). Interestingly, Leslie and Arnott (1978) observed the same molecular conformation in an orthorhombic crystal form of poly(2′-OMe-C) where there can be no hydroxyl–hydroxyl hydrogen bonds.

EXTERNAL FACTORS AFFECTING STRUCTURAL TRANSITIONS IN DNA AND RNA DUPLEXES

The $A \rightarrow B$ transition was first observed in fibres of typical DNAs when the ambient relative humidity was increased from 75 per cent to 92 per cent (Franklin and Gosling, 1953). Later it was recognised that the presence of extraneous ions (such as about 3 per cent NaCl) was also necessary and that with fibres containing twice as much salt only *B*-DNA would be observed (Cooper and Hamilton, 1966). The *C*-DNA conformation is sometimes observed under intermediate conditions of salt content and relative humidity (Marvin *et al.*, 1961; Brahms *et al.*, 1973; Arnott and Selsing, 1975; Leslie *et al.*, 1980).

The type of cation present can also affect the transition points so that with NaDNA *A*, *C*, and *B* forms are all observed but with LiDNA only *C* and *B* forms. Transitions analogous to the $B \rightarrow C \rightarrow A$ structural changes in fibres apparently occur in aqueous solutions to which alcohols have been added (Brahms and Mommaerts, 1964; Malenkov *et al.*, 1975). Therefore, it may be concluded that the *B* conformation is favoured when the environment is most polar, most of the other conformations as it becomes less so.

By contrast RNA duplexes and DNA–RNA hybrid duplexes are relatively little affected by changes in the polarity of the environment. The minor transition $A \rightarrow A'$ can be obtained in the same circumstances which would cause a $A \rightarrow B$ or $D \rightarrow B$ transition in a DNA duplex (Arnott *et al.*, 1968; Banerjee *et al.*, 1980).

INTERNAL FACTORS AFFECTING DNA DUPLEX STRUCTURES

Internal factors are defined as nucleotide composition, sequence and repetitiveness of sequence. We have already seen that ribonucleotides constrain duplexes not to be *B*. To determine the effects of the other factors on DNA duplexes Leslie *et al.*

Table 4.4a Different polydinucleotide and polytrinucleotide duplexes with the repetitious base sequences indicated exhibit the allomorphic variants shown in italics

AT	AC	AG	GC
TA	TG	TC	CG
B, D, (A)	*S, B, A*	*B, C″*	*S, B, A*
AAA	AAT	AAC	AAG
TTT	TTA	TTG	TTC
B′	*B, D*	*B, A*	*?*
	AGT	AGC	
	TCA	TCG	
	B, C, A	*B, C′, A*	
	GAT	GAC	
	CTA	CTG	
	B, A	*?*	
AGG	GGT	GGC	GGG
TCC	CCA	CCG	CCC
?	*B, C′, A*	*?*	*B, A*

Table 4.4b Polymorphism in polydinucleotide and polytrinucleotide duplexes where G has been replaced by I

	AC	AI	IC
	CI	CT	CI
	B, C, D	*B′*	*B, D*
		AAC	AAI
		TT I	TTC
		?	*?*
	AIT	AIC	
	TCA	TCI	
	B, C, D, (E)	*B, D*	
	IAT	IAC	
	CTA	CTI	
	?	*?*	
AI I	IIT	I IC	III
ICC	ICC	ICC	CCC
?	*B, E*	*?*	*(B), B′*

(1980) undertook a systematic study of the four synthetic DNAs with repeating dinucleotide sequences and eight of the twelve DNAs with repeating trinucleotide sequences. With these DNAs it is possible to explore extremes of base composition (from 100 per cent AT in three instances to 100 per cent GC in two instances) and sequence from homopolymer chains to alternating purine/pyrimidine nucleotide sequences. Leslie *et al.* (1980) also considered a number of analogues where G had been replaced by I (hypoxanthine) and discovered that in essence I:C = A:T ≠ G:C.

The results of both studies are summarised in tables 4.4a and 4.4b. A number of new structures were discovered. Some, like *B'*, *C'*, *C"*, are merely minor variants of classical structures. Others like *S, D, E* have quite novel features. It would appear that it is nucleotide sequence rather than composition or repetitiveness of sequence which is the important factor in determining structural behaviour. This is most evident with *S*-DNA which can accommodate only alternating purine/pyrimidine base sequences and these only when the purine is on the *syn* nucleoside but it is also evident in the quite different properties of isocompositional DNAs like poly d(A)·poly d(T) and poly d(AT)·poly d(AT), poly d(G)·poly d(C) and poly d(GC)· poly d(GC), poly d(AC)·poly d(GT) and poly d(AG)·poly d(CT).

The classical *B*-DNA structure is observed with all DNAs except poly d(A)·poly d(T) and poly d(AI)·poly d(CT). With these specimens, however, one has to avoid fibres with retained salt since this encourages disproportionation to forms like poly d(T)·poly d(A)·poly d(T). It is almost certain that the *B* form is available to such sequences but that it cannot be trapped under our experimental conditions.

The classical *A*-DNA form is rare or absent for sequences with no G (and also for poly d(AG)·poly d(CT)). Its occasional observation with poly d(AT)·poly d(AT) indicates that this is a preference rather than an absolute constraint. Indeed, it should always be remembered that when compared to the variety of environmental conditions that may be experienced by DNA molecules *in vivo*, the parameters which we have been able to control – sodium or lithium cations, relative humidity, amount of retained salt – are very limited. There is no way of assuredly mimicking in hydrated fibres the effect of intimate protein–DNA interaction which takes place in cells. It might be expected that some protein binding would reduce the polarity of the DNA environment and tend to produce the structural types observed in fibres with low quantities of retained salt. In these circumstances, some sequences, especially those rich in A:T pairs, may be more likely than others to adopt non-classical DNA conformations. However, there are no stereochemical reasons why any base pair need be excluded from these new structures and, therefore, it is possible that *any* DNA could adopt these conformations as a result of appropriate interactions with a bound protein. Certainly, there are proteins with sufficient concentrations of positive charges to produce the 'high salt' conditions needed for the transition to the left-handed *S* form.

POSSIBLE BIOLOGICAL ROLES FOR DIFFERENT DNA SECONDARY STRUCTURES

Although native DNA might be envisioned with many different, sequence-dependent secondary structures, the evidence is in favour of it having a relatively monotonous secondary structure something like *B*-DNA. There may be rare discontinuities but for the most part a local sequence on a DNA would be able to display its special structural preference only when isolated from the effects of other sequences and when in a special polar environment such as might be provided *in vivo* by protein binding sites.

The importance of the considerable polymorphism observed in fibre specimens of DNA is manyfold. Clearly, the effects of particular sequences can be amplified by their being present, or being induced to be present, in peculiar secondary structures which would enhance the availability of certain base functional groups and diminish the accessibility of others.

The discovery of secondary structures notably overwound or grossly underwound compared with *B*-DNA is important in connection with the development of higher order coiling in DNA tertiary structure: overwound and underwound secondary structures would have to be present in left-handed and right-handed supercoils. Moreover, local regions of well-packed DNAs might be expected to be topologically quite constrained so that the net winding of a supercoiled DNA might have to be about the same as a relaxed DNA. Compensating blocks of 'high wound' or 'low wound' secondary structures might have to be provided locally. We can now envisage what these might be and what sequences would be more hospitable for them. Conversely the induction of special secondary structures in such sequences could trigger tertiary structure formation in a fairly specific way.

The availability of high and low wound DNA forms also simplifies the unwinding problem – conceptually at least – and may help to render moot many topological problems thought to be associated with some DNA activities. The known DNA conformations that are substantially overwound or underwound compared with *B*-DNA could be used to store or shunt locally the rotations needed to wind or unwind a double helix. For example, a segment of left-handed DNA, maintained in a region where the preferred sequences were sufficiently common, could be a store of negative windings. In appropriate conditions these could be used to compensate positive windings and produce a region of melted DNA available for polynucleotide synthesis. Conversely, regions of unwound DNA could condense to give contiguous segments of left-handed and right-handed helical DNA separated by as few as two nucleotides. Such a structure might be a manageable substrate for topological isomerisation. Alternatively, the left-handed segment would form a package of negative windings which could be translocated easily even to a distant site.

REFERENCES

Arnott, S. (1975). *Nucl. Acids Res.*, **2**, 1493.

Arnott, S. and Bond, P. J. (1973a). *Nature, New Biology*, **244**, 99.

Arnott, S. and Bond, P. J. (1973b). *Science*, **181**, 58.

Arnott, S., Bond, P. J. and Chandrasekaran, R. (1980). *Nature, Lond.* (submitted).

Arnott, S. and Chandrasekaran, R. (1980a). *J. molec. Biol.* (submitted).

Arnott, S. and Chandrasekaran, R. (1980b). *Nature, Lond.* (submitted).

Arnott, S., Chandrasekaran, R. and Leslie, A. G. W. (1976). *J. molec. Biol.*, **106**, 735.

Arnott, S., Chandrasekaran, R. and Marttila, C. (1974). *Biochem. J.*, **141**, 537.

Arnott, S. and Hukins, D. W. L. (1972a). *Biochem. biophys. Res. Commun.*, **47**, 1504.

Arnott, S. and Hukins, D. W. L. (1972b). *Biochem. J.*, **130**, 453.

Arnott, S. and Hukins, D. W. L. (1973). *J. molec. Biol.*, **81**, 93.
Arnott, S. and Selsing, E. (1974). *J. molec. Biol.*, **88**, 509.
Arnott, S. and Selsing, E. (1975). *J. molec. Biol.*, **98**, 265.
Arnott, S. and Wonacott, A. J. (1966). *Polymer*, **7**, 157–166.
Arnott, S., Wilkins, M. H. F., Fuller, W. and Langridge, R. (1967). *J. molec. Biol.*, **27**, 535.
Arnott, S., Fuller, W., Hodgson, A. and Prutton, I. (1968). *Nature, Lond.*, **220**, 561.
Arnott, S., Hukins, D. W. L., Dover, S. D., Fuller, W. and Hodgson, A. R. (1973). *J. molec. Biol.*, **81**, 107.
Arnott, S., Chandrasekaran, R., Hukins, D. W. L., Smith, P. J. C. and Watts, L. (1974). *J. molec. Biol.*, **88**, 23.
Arnott, S., Chandrasekaran, R., Birdsall, D. L., Leslie, A. G. W. and Ratliff, R. L. (1980*a*). *Nature, Lond.*, **283**, 743.
Arnott, S., Chandrasekaran, R., Day, W. A., Puigjaner, L. and Watts, L. (1980*b*). *J. molec. Biol.* (submitted).
Arnott, S., Chandrasekaran, R., Leslie, A. G. W., Puigjaner, L. and Saenger, W. (1980*c*). *J. molec. Biol.* (submitted).
Banerjee, A., Arnott, S., Leslie, A. G. W. and Selsing, E. (1980). *J. molec. Biol.*, (submitted).
Beetz, Jr., C. P., Ascarelli, G. and Arnott, S. (1979). *Biophys. J.*, **28**, 15.
Bond, P. J., Langridge, R., Jennett, K. W. and Lippard, S. J. (1975). *Proc. natn. Acad. Sci. U.S.A.*, **72**, 4825.
Brahms, J. and Mommaerts, W. F. H. (1964). *J. molec. Biol.*, **10**, 73.
Brahms, J., Pilet, J., Lan, T.-T.P. and Hill, L. R. (1973). *Proc. natn. Acad. Sci. U.S.A.*, **70**, 3352.
Chou, C. H., Thomas, G. J., Jr., Arnott, S. and Smith, P. J. C. (1977). *Nucl. Acids Res.*, **4**, 2407.
Cooper, P. J. and Hamilton, L. D. (1966). *J. molec. Biol.*, **16**, 562.
Crick, F. H. C. and Watson, J. D. (1954). *Proc. R. Soc. A.*, **223**, 80.
Franklin, R. E. and Gosling, R. G. (1953). *Acta crystallogr.*, **6**, 673.
Furberg, S. (1950). *Acta crystallogr.*, **3**, 325.
Gray, D. M. and Tinoco, Jr., I. (1970). *Biopolymers*, **9**, 223–244.
Ladner, J. E., Jack, A., Robertus, J. D., Brown, R. S., Rhodes, D., Clark, B. F. C. and Klug, A. (1975). *Proc. natn. Acad. Sci. U.S.A.*, **72**, 4414.
Langridge, R. and Gomatos, P. J. (1963). *Science*, **141**, 694.
Langridge, R., Marvin, D. A., Seeds, W. E., Wilson, H. R., Hooper, C. W., Wilkins, M. H. F. and Hamilton, L. D. (1960). *J. molec. Biol.*, **2**, 38–64.
Langridge, R. and Rich, A. (1963). *Nature, Lond.*, **198**, 725.
Langridge, R., Seeds, W. E., Wilson, H. R., Hooper, C. W., Wilkins, M. H. F. and Hamilton, L. D. (1957). *J. biophys. biochem. Cytol.*, **3**, 767.
Leslie, A. G. W. and Arnott, S. (1978). *J. molec. Biol.*, **119**, 399.
Leslie, A. G. W., Arnott, S., Chandrasekaran, R. and Ratliff, R. L. (1980). *J. molec. Biol.* (in press).
Malenkov, G., Minchenkova, L., Minyat, E., Schyolkina, A. and Ivanov, V. (1975). *FEBS Lett.*, **51**, 38.
Marvin, D. A., Spencer, M., Wilkins, M. H. F. and Hamilton, L. D. (1961). *J. molec. Biol.*, **3**, 547.
Milman, G., Langridge, R. and Chamberlain, M. J. (1966). *Proc. natn. Acad. Sci. U.S.A.*, **57**, 1804.
Mitsui, Y., Langridge, R., Shortle, B. E., Cantor, C. R., Grant, R. C., Kodama, M. and Wells, R. D. (1976). *Nature, Lond.*, **228**, 1166.
O'Brien, E. J. and McEwan, A. W. (1970). *J. molec. Biol.*, **48**, 243.
Patel, D. J., Canuel, L. L. and Pohl, F. M. (1979). *Proc. natn. Acad. Sci. U.S.A.*, **76** 2508.

Pilet, J. and Brahms, J. (1972). *Nature: New Biology*, **236**, 99.
Pilet, J., Rottman, F. and Brahms, J. (1973). *Biochem. biophys. Res. Commun.*, 52, 517.
Rich, A. (1958). *Biochim. biophys. Acta*, **29**, 502.
Smith, P. J. C. and Arnott, S. (1978). *Acta crystallogr.*, A**34**, 3.
Wang, A. H.-J., Quigley, G. J., Kolpak, F. J., Crawford, J. L., van Boom, J. H., van der Marel, G. and Rich, A. (1979). *Nature, Lond.*, **282**, 680.
Watson, J. D. and Crick, F. H. C. (1953). *Nature, Lond.*, **171**, 737.
Zimmerman, S. R., Cohen, G. H. and Davies, D. R. (1975). *J. molec. Biol.*, **92**, 181.

5

Transfer RNA: crystal structures

Sung-Hou Kim

INTRODUCTION

Some of the most interesting aspects of transfer RNA (tRNA) are the extremely diverse and essential roles it plays in living cells. The best understood role is its participation in the decoding process of genetic codes during ribosome-mediated protein synthesis (for a review see Haselkorn and Rothman-Denes, 1973).

Many other roles (for a review see LaRossa and Söll, 1978) have been identified with tRNA, for example, in cell-wall synthesis in bacteria through a special tRNA species, in amino acid transport, in regulation of certain amino acid biosyntheses, in inhibition of enzyme activity, in RNA-dependent DNA synthesis.

Besides tRNA being a functionally interesting molecule, its manageable size makes the molecule suitable for x-ray crystallographic study. Ever since the double helical model of DNA was proposed, the importance of the model in understanding the function of DNA has been well recognised. Furthermore, the needs for *detailed* structural information on nucleic acid in general have been increasing rapidly, especially in recent years when understanding of the protein-nucleic acid interaction at the structural level demands the conformational details of the molecules involved in free as well as in the interacting states.

Since the crystal structure of yeast phenylalanine tRNA, tRNA[Phe], is to date the best-refined nucleic acid structure determined by single crystal x-ray diffraction methods (the best technique for structural details at present), it can serve as the basis for the understanding not only of the structure-function relationship of tRNA but also as a source for conformational details on secondary and tertiary structure of nucleic acids in general.

There are several review articles on the three-dimensional structure of yeast tRNA[Phe] emphasising different aspects (Clark, 1977; Kim, 1979; Rich and RajBhandary, 1976). This chapter provides a summary of various structural features of tRNA[Phe], indicates any significant differences between this tRNA and others currently being studied, and points out the structural features of functional interest.

PRIMARY AND SECONDARY STRUCTURE OF tRNA

Cloverleaf model

Utilising chromatography on DEAE-Sephadex, Holley *et al.* (1965) established the first nucleotide sequence (minor corrections of the sequence have since been made) of an RNA, yeast tRNA^{A1a}, and proposed the well-known 'cloverleaf' pattern for the secondary structure of tRNA. Since then, the two-dimensional electrophoresis technique has greatly accelerated tRNA sequencing studies. Over 90 different tRNAs have been sequenced, and all can be arranged into the cloverleaf secondary structure pattern (for a convenient compilation, see Sprinzl *et al.*, 1978). A generalised cloverleaf pattern is shown in figure 5.1 together with the 'L' pattern, which is a rearrangement of the cloverleaf pattern to simulate the three-dimensional structure of tRNA.

The common features of the secondary structure are as follows. Starting from the 5' end (see figure 5.1), all the tRNAs have 7 base pairs in the amino acid stem (AA stem), 3 or 4 base pairs in the dihydrouracil stem (D stem), 5 base pairs in the anticodon stem (AC stem), and 5 base pairs in the TψC stem (T stem). The base pairs in the stems are of the Watson-Crick type, with occasional exceptions. There are always 2 nucleotides between the AA and D stems, 1 between the D and AC stems, and no nucleotide between the AA and T stems. There are always 7 nucleotides in the AC and T loops, 7-10 nucleotides in the D loop and 4-21 nucleotides in the variable arm (V arm).

Among all tRNA sequences so far determined, about 50 per cent of them belong to a subclass, commonly called class I or D_4V_5. All tRNAs in this class contain 4 base pairs in the D stem and 5 bases in the V loop. Yeast tRNA^{Phe} belongs to this class.

Conserved and semiconserved bases

As shown in figure 5.1 certain bases are conserved among all elongation tRNAs: U between the AA and D stems; A at the beginning of the D loop; two Gs in the D loop; U in the AC loop; a GC base pair in the T stem; the T-ψ-C sequence and A in the T loop and the C-C-A sequence at the 3' end.

Similarly, there are semiconserved bases. (The term semiconserved is used when a position in the sequence is always occupied by the same type of base and indicated by R (for purine) or Y (for pyrimidine) in figure 5.1.) For example, the base after the sequence T-ψ-C in the T loop is always either G or A, that is, a purine. Most of these are conserved and the semiconserved bases are located in the D arm (arm refers to a stem and a loop) and the T loop.

It may be safely said that these residues must play important roles in providing either the recognition surfaces common to all tRNAs or the basic architectural framework to build up the three-dimensional structure of tRNA, or both. X-ray crystallographic studies on yeast tRNA^{Phe} reveal that the majority of these conserved and semiconserved bases are involved in forming the tertiary structure of this tRNA (see p. 108).

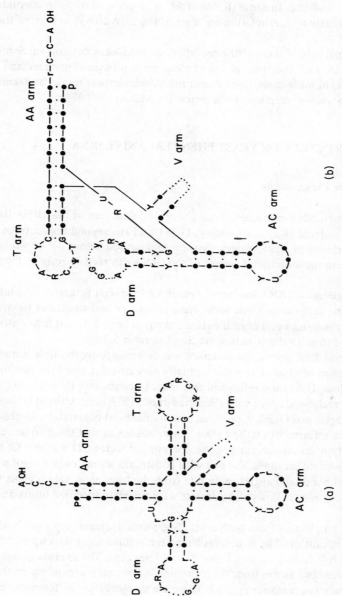

Figure 5.1 Generalised representation of tRNA in (a) the 'cloverleaf' form and (b) the 'L' form. The bases common to all tRNAs (participating in peptide elongation) are indicated. Other symbols are R for a purine nucleoside and Y for a pyrimidine nucleoside in *all* tRNAs, r for a purine nucleoside and y for a pyrimidine nucleoside in *most* tRNAs, p for (5′) terminal phosphate, and OH for (3′) terminal hydroxyl. The dotted lines are the regions in the chain where the number of nucleotides varies among tRNAs.

Modified nucleosides

One of the characteristic aspects of tRNAs is that a significant portion of the nucleosides are modified. In some tRNAs close to 16 per cent of the nucleosides are modified, mostly by methylation on bases or the 2′ hydroxyl oxygen of the riboses.

The functional role of most of the modified nucleosides is not known. Some of these (T and ψ in the T loop) are invariant elements in all except mammalian initiator tRNAs. In most cases, these minor nucleosides appear not to be essential for the amino-acylation reaction (for a review see Nishimura, 1978).

CRYSTAL STRUCTURE OF YEAST PHENYLALANINE tRNA

Crystallographic backgrounds

Experimental methods used to determine the crystal structure of this tRNA have been reviewed in detail (Kim and Quigley, 1978), and are beyond the scope of this chapter. The primary crystallographic data are summarised below. Those who are interested only in the structural features of the tRNA can turn directly to the next section.

Yeast phenylalanine tRNA has been crystallised in several forms, of which two, an orthorhombic and a monoclinic form, have yielded crystal structures so far. The crystallisation conditions and crystal data are given in table 5.1, and heavy atom derivatives used to solve the structure are listed in table 5.2.

For both crystalline forms, the structure was determined by multiple isomorphous replacement methods (for a conceptually easy description of the method, see Holmes and Blow, 1965) and refined in real space to begin with (Stout *et al.*, 1978), then finally in reciprocal space with (Sussman *et al.*, 1978) and without (Hingerty *et al.*, 1978; Teeter and Quigley, personal communication) constraints. In general, reciprocal space refinement method gives 'better' results in that the difference electron density maps are cleaner and the discrepancy factor (R) is lower. Of the reciprocal refinement methods, the one with constraints allows only a small number of chemical bonds and angles to deviate from the 'canonical' values while the one without constraints allows all to deviate within certain specified limits during refinements.

Diffraction amplitudes from both crystalline forms decrease very rapidly after around 2.7 Å resolution. The structures have been refined using data up to 2.5 Å resolutions with R factors ranging from 20 to 30 per cent. The extents of distortion of bond distances and angles from the 'canonical' values vary depending on the particular refinement methods applied. There are no *significant* differences among the structures refined by four different research groups. Some of the minor differences will be pointed out later in the text.

Table 5.1 Crystallisation conditions and crystal data of yeast tRNA^{Phe}

Initial tRNA concentration (mg ml^{-1})	Buffer	pH	Other	Precipient	Crystallisation method	Temperature (°C)	Period	Space group	Cell dimensions (Å)	References
4	Na cacodylate 10 mM	6.0	MgCl$_2$ 10 mM, spermine 4HCl 1 mM	Isopropanol 10% (v/v)	Vapour diffusion	4	3 days ~2 weeks	P2$_1$2$_1$2$_1$	33, 56, 161	[a]
15	Na cacodylate 40 mM	6.0	MgCl$_2$ 40 mM, spermine 4 HCl 4 mM	Isopropanol 8% (v/v)	Vapour diffusion	4	1 day ~ 2 weeks	P2$_1$2$_1$2$_1$	33, 56, 161	[a]
3.4	Na cacodylate 10 mM	6.0	MgCl$_2$ 10 mM, spermine HCl 1 mM	MPD 10–13% (v/v)	Vapour diffusion	7	~2 weeks	P2$_1$	33, 56, 63 $\alpha = 90°$	[b]
~2	K cacodylate 10 mM	7.0	MgCl$_2$ 5–15 mM, spermine HCl 1–3 mM	Dioxane 10–20% (v/v)	Vapour diffusion	4	~3 days	P2$_1$	33, 56, 63 $\alpha = 90°$	[c]

[a] Kim et al. (1971)
[b] Ichikawa and Sundaralingam (1972)
[c] Ladner et al. (1972)

Table 5.2   Heavy-atom derivatives and their soaking conditions
for yeast phenylalanine tRNA crystals

Compound	Concentration	Soak time	Reference
trans-Diaminodichloroplatinate *trans*-PtCl$_2$(NH$_3$)$_2$	0.2	1–2 days	a
Potassium tetrachloroplatinate K$_2$PtCl$_4$	1.0	$>$ 2 days	b
Potassium tetracyanoplatinate K$_2$Pt(CN)$_4$	1.0	$\sim$ 1 week	b
Sodium aurous cyanide NaAu(CN)$_2$	1.0	$\sim$ 1 week	b
Hydroxymercurihydroquinone-O, O-diacetate	1.0	1–2 days	a
Samarium acetate Sm(CH$_2$COO)$_3$	1.0	2–7 days	b
	0.2–0.5	2–3 days	a
Lutetium acetate Lu(CH$_3$COO)$_3$	1.0	2–7 days	b
Lutetium chloride LuCl$_3$	0.2–0.5	2–3 days	a
Praseodymium nitrate Pr(NO$_3$)$_3$	1.0	2–7 days	b
Acetates of Eu, Tb, Dy, Gd	1.0	2–7 days	b
Bis(pyridine) osmate [Py]$_2$OsO$_3$	K$_2$OsO$_4$ 2 mM, pyridine 62 mM	2 weeks	c
Potassium osmate K$_2$OsO$_4$	1.0	$\sim$ 2 weeks	b
Bis(pyridine) osmate·ATP	solid added	$\sim$ 2 weeks	b
	0.6	5 days	a
Bis(pyridine) osmate complex of CTP, UTP, AMP, GMP	solid added	$\sim$ 2 weeks	b

[a]Jack, Ladner and Klug (1976).
[b]Suddath *et al.* (1974) and unpublished results.
[c]Stout *et al.* (1978).

Overall structure

The crystal structure of yeast tRNA[Phe] has an overall shape of the letter L (Kim
et al., 1973). The polynucleotide backbone of this model is folded so that the
amino acid stem and T stem form one continuous double helical arm, and the D
stem and the anticodon stem form the other long double helical arm of the L. Each
extension of the L is about 60 Å long with a diameter of about 20 Å and the
distance between the two extremes is about 80 Å. Each stem is an antiparallel,
right-handed double helix similar to *A*-RNA. The polynucleotide backbone structure of the molecule is shown in figure 5.2 and a space filling model is shown in
figure 5.3 to provide a more 'realistic' and spatial appearance of this molecule.

The 3′-end, where peptide elongation occurs, is at one extreme of the molecule,
while the anticodon triplet, which recognises the codon on the messenger RNA is at
the other extreme. The T loop, which has largely conserved base sequence in all
elongation tRNAs and is implicated as a ribosomal RNA recognition site (for a
review see Erdmann, 1978), appears at the corner of the L. Thus these three functionally important sites are maximally separated, which may help minimise mutual
interference among them and their corresponding sites on the ribosomes.

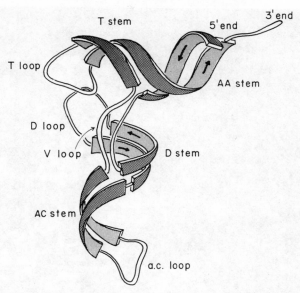

Figure 5.2 The flow of the backbone of the yeast tRNA[Phe] structure. The 5' end to 3' end direction is indicated by arrows.

Irregular details of double helical stems

Four double helical stems predicted from the cloverleaf model have been found in the crystal structure. All four stems have a conformation similar to A-RNA in that they have a shallow groove and a deep groove; the base pairs are considerably tilted from the helical axes (ranging from 13° in anticodon stem to 20° in D stem on the average). Various helical parameters calculated, based on the refined atomic co-ordinate of this tRNA in both crystal forms (Hingerty *et al.*, 1978; Holbrook *et al.*, 1978; Sussman *et al.*, 1978), are given in table 5.3. The helix rotation is usually larger and the displacement shorter than those of A-RNA. Among the four stems, the D stem differs most from A-RNA. This is probably due to the extensive tertiary interactions made by this stem with other parts of the molecule. The angle between the two helical axes of the acceptor and T stems is 14°, and the corresponding angle between the two helical axes of the D stem and the anticodon stem is 24°. The angle between the two arms of L is about 92°. These are listed in table 5.4.

One interesting feature of these four stems is the apparent high irregularity of their helicity. As is clear from the large estimated standard deviation associated with helical parameters (for the orthorhombic form) (table 5.3) all four helices have considerable irregularities. It is not clear, however, whether these irregularities reflect the sequence dependence or an inherent flexibility of the backbone conformation locked into one form in this particular crystal form. Notice also that average twist angles between two bases that are base paired are considerably greater than that of A-RNA.

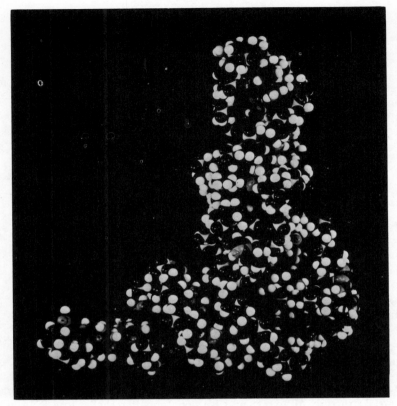

Figure 5.3   A space filling model of yeast tRNA^{Phe} structure.

Table 5.3   Helical parameters of the stems

Stem	Number of residues per turn		Helix rotation[a] (°)		Displacements[b] (Å)		Average tilt angle[c] (°)		Average twist angle[d] (°)	
	(I)[e]	(II)[f]	(I)	(II)	(I)	(II)	(I)	(II)	(I)	(II)
Amino acid stem	10.8	11.0	33(2)	33	2.5(0.3)	2.6	16(2)	20	11(6)	5
D stem	10.3	11.2	35(4)	32	2.4(0.3)	2.6	21(3)	23	6(2)	7
T stem	10.6	10.9	34(2)	33	2.5(0.2)	2.6	16(3)	16	10(7)	4
Anticodon stem	11.0	11.0	33(4)	33	2.7(0.4)	2.5	13(3)	20	5(2)	8
A-RNA	11.0		32.7		2.8		13		2.2	

[a]Angle to be rotated around the helix axis to bring one nucleotide to the next on the same strand.

[b]Translation per base pair along the helix axis.

[c]Angle between normal to base and helix axis.

[d]Angle between two bases that are base paired.

[e]Holbrook *et al.* (1978).

[f]Hingerty *et al.* (1978).

Table 5.4 Interhelical angles[a]

Amino acid stem	−D stem	107°
Amino acid stem	−anticodon stem	97°
Amino acid stem	−T stem	14°
D stem	−Anticodon stem	24°
D stem	−T stem	70°
Anticodon stem	−T stem	85°
(amino acid−T stem)	−(D − anticodon stem)	92°

[a]Calculated based on the least squares helical vectors obtained using C1′, P and N9 of purine or N1 of pyrimidine.

Extensive base stacking

Although only 55 per cent of the bases in this molecule are in the double helical stems, the three-dimensional structure in both crystal forms reveals that 93 per cent are stacked; all except five bases (D16, D17, G20, U47, and A76) are stacked as schematically shown in figure 5.4 (centre). Such extensive base stacking, including bases in the loops, must be one of the major stabilising forces, and is likely to be a universal feature of *all* free nucleic acids.

Tertiary base pairs of non-Watson-Crick type

There are nine tertiary base pairs in the crystal structure of this tRNA as shown schematically in figure 5.4 (Hingerty *et al.*, 1978) assigned one more, D16·C60, which forms a distorted hydrogen bond). Two remarkable points are that *none* of these tertiary base pairs are of the Watson-Crick type except the G19. C56 pair, and that most of these involve bases that are conserved or semiconserved in all tRNAs (figure 5.5a), thus giving credence to the hypothesis that all tRNAs have the same general structural frame (Kim *et al.*, 1974; Klug *et al.*, 1974). All of these are located at the intersection of the two arms of the L, and are essential for maintaining the L-shaped basic frame (figure 5.5b). Location of conserved and semiconserved bases and tetiary base pairs are indicated in cloverleaf, L-diagram, and crystal structure in figure 5.5. The details of regions containing tertiary base pairs are shown in figure 5.6. There are no discrepancies between the four research groups in the assignments of tertiary base pairs except D16·C60 mentioned above.

Assignments of secondary and tertiary hydrogen bonds *between* bases in electron density maps are relatively easy because the number of relative locations of the paired bases are limited and predictable. However, assignment of hydrogen bonds involving the backbone is much more difficult and arbitrary. This is even more so for the backbone–backbone hydrogen bonds. Therefore, assignments of this class of hydrogen bonds vary considerably between the four research groups and should be considered as tentative at the present time.

Twelve hydrogen bonds of this category assigned for the tRNA in an orthorhombic crystal form are shown in figure 5.6. Of these, five involve O2′ hydroxyl groups.

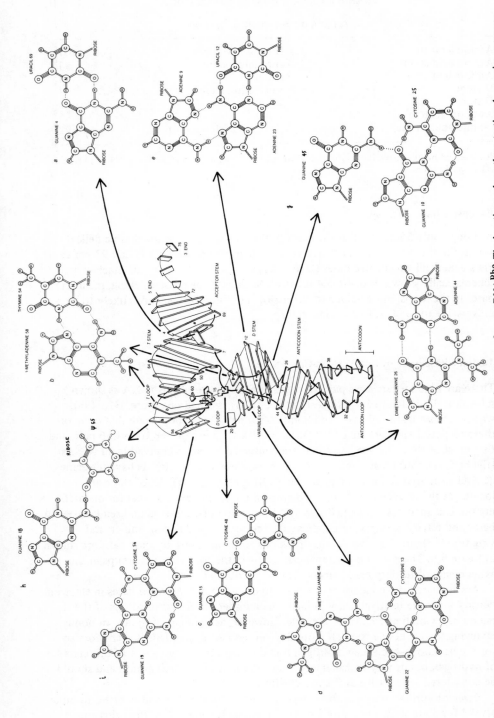

Figure 5.4 Nine tertiary base pairs and one G·U base pair in the crystal structure of yeast tRNA[Phe]. The locations where these tertiary base pairs occur are indicated. Notice that all tertiary base pairs are different from Watson-Crick base pair type, and that all are loacted near the corner of the L.

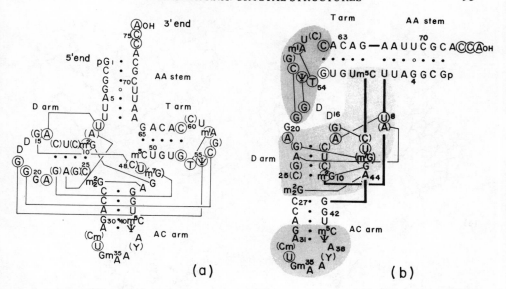

Figure 5.5 Nucleotide sequence of yeast tRNA^Phe in (a) cloverleaf and (b) L arrangement. Conserved bases are in circles and semiconserved in parentheses, for $D_4 V_5$ class tRNAs (tRNAs with four base pairs in the D stem and 5 bases in the V loop). The bases that form base–base tertiary H-bonds are connected by thin lines. Notice that most of the conserved and the semiconserved bases are localised to the middle of the L, and are involved in forming the tertiary hydrogen bonds. The details of the actual structure for the shaded regions in (b) are shown in figure 5.6.

Extensive utilisation of O2′ hydroxyl groups to form tertiary hydrogen bonds may be the reason why RNA rather than DNA is used for 'structural' nucleic acids such as tRNA and ribosomal RNAs.

Partially flexible anticodon loop

The anticodon loop is rather free from tertiary hydrogen bonds. Five bases including the anticodon triplet are stacked on one side of the loop leaving only two bases on the other side (figures 5.4, 5.5b). A detailed examination of anticodon triplet conformation reveals that, in the orthorhombic crystal, it is very similar to that of the A-RNA helix (Holbrook et al., 1978), whereas, in the monoclinic crystal, it assumes a much tighter conformation, that is, the turn angle per residue is much greater than that of A-RNA (Jack et al., 1976). An additional difference is that in the orthorhombic crystal, the anticodon loop is stabilised by a magnesium hydrate ion; Mg^{2+} forms a direct coordination bond to a phosphate oxygen of residue 37, and the water molecules (coordinated to Mg^{2+} ion) form several hydrogen bonds to nearby bases in this loop (Holbrook et al., 1977; Quigley et al., 1978). But in the monoclinic crystal no magnesium ion was found in this loop (Hingerty et al., 1978). These two differences can be understood if it is assumed that the anticodon loop is

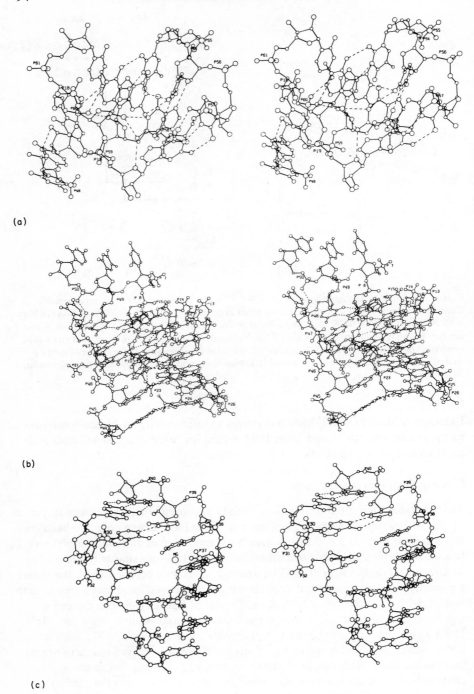

(a)

(b)

(c)

partially flexible and the differences merely reflect the difference of crystal packing environments in two crystal forms. The partial flexibility of the anticodon loop is also manifested in its higher thermal motion compared to the other parts of the molecule (see section on Molecular flexibility).

Tightly bound magnesium ions

It has been known that specific-site bound magnesium ions help to stabilise the functional conformation of transfer RNAs (Fresco *et al.*, 1966). This subject has been recently reexamined (Stein and Crothers, 1976). There are two magnesium hydrate binding sites in the D loop, one in the sharp turn formed by the residues 8–12, and one in the anticodon loop. The last one was not found in the monoclinic crystal (Jack *et al.*, 1977). The assignment and environment of these magnesium binding sites are listed in table 5.5 and shown in figure 5.7. From these locations it is clear that their functional role is to stabilise the loops and sharp turns of the tRNA structure in order to maintain its functional conformation.

In addition, there are many weak electron density peaks that can be interpreted as weak magnesium binding sites with much less confidence. Positions of randomly bound magnesium ions are not obtainable by x-ray crystallographic methods.

Spermine in deep groove

Polyamines such as spermine are known to stabilise the functional conformation of tRNA structure. One spermine molecule has been located in the deep groove of the long double helix formed by the D stem and the anticodon stem in both crystal forms (Hingerty *et al.*, 1978; Holbrook *et al.*, 1978; Quigley *et al.*, 1978). The second one was assigned in the deep groove of the double helix formed by the amino acid stem and the T stem by Holbrook *et al.* (1978), and between the variable loop and the residue 10 by Quigley *et al.* (1978). Although the one in the D anticodon stem is easily recognisable in the difference electron density maps, the assignments of the second one are not reliable. In any event, one of the roles spermine plays is to stabilise the RNA double helix by binding to its deep (major) groove. These are also shown in figure 5.7 and table 5.5.

The water-accessible surface of tRNA

There are many electron density peaks interpretable as bound water molecules (based on distance criteria), mostly along the deep grooves of double helical stems, with some on shallow grooves. The positional accuracies as well as assignment of

Figure 5.6 Stereo views of the three regions of yeast tRNA[Phe] indicated in figure 5.5b. Magnesium ions are shown with the largest diameter, then phosphorus, oxygen, nitrogen and carbon in decreasing order of size. All the hydrogen bonds are indicated by broken lines. (a) Outer corner of the L, (b) inner vertex of the L, (c) the anticodon loop and a part of the anticodon stem.

Table 5.5 tRNA-ligands interaction[d]

Assignment	Location	Direct coordination to	H-Bonded to	References
$[Mg(H_2O)_5]^{2+}$	D loop	Phosphate oxygen of G19	Phosphate oxygen (G19) Bases (G20, U59, C60)	a, b, c
				a, b
				a, c
			Bases (G20, U59, C56)	b
$[Mg(H_2O)_4]^{2+}$	D loop	Phosphate oxygens of G20, A21		a, b, c
$[Mg(H_2O)_6]^{2+}$	8–12 turn		Phosphate oxygens (U8, A9, C11, U12)	a, b, c
$[Mg(H_2O)_5]^{2+}$	Anticodon loop	Phosphate oxygen of Y37	Bases (C32, Y37, A38, ψ39)	a, b
Spermine	Wide grooves D stem and part of anticodon stem		PO$_4$ (23, 24, 25, 42, 44)	a, b, c
Spermine	Wide grooves of amino acid and T stems			a
Spermine	Variable loop-G10		PO$_4$ (10, 46, 47, 45)	b

[a](Holbrook et al., 1978).
[b](Quigley et al., 1978).
[c](Hingerty et al., 1978).
[d]Ligand binding between two tRNAs in crystals is not listed.

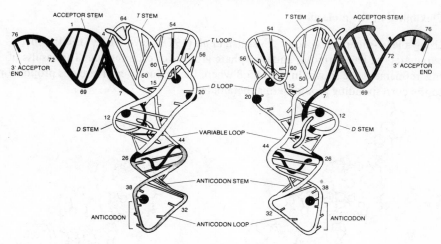

Figure 5.7 Two views of the folding of backbone of yeast tRNA^{Phe} are shown diagramatically. The sugar-phosphate backbone of the molecule is represented as a coiled tube with the cross-rungs showing the nucleotide base pairs in the stem regions. The short rungs indicate bases that are not involved in base–base H-bonding. Dark circles represent four magnesium hydrate ions and solid "spaghetti" represents spermine molecule found with more certainty and open "spaghetti" represent the spermine positions determined with less certainty (see table 5.5).

these peaks are quite poor and unreliable in both crystal forms. However, given the three-dimensional structure, one can calculate the water-accessible area.

The method of calculation was described by Lee and Richards (1971) in their calculation of polypeptide surface areas. A spherical 'probe' or solvent molecule of radius r_w free just to touch but not penetrate the van der Waals surface of the examined molecule. The closed surface defined by all possible loci for the centre of the probe is described as the accessible surface of the molecule.

Based on these calculations, the following three interesting results have been obtained (Alden and Kim, 1979).

(1) The folded tRNA molecule has only one-third of the total water-accessible surface areas of extended, unfolded tRNA, and most of the buried surfaces are those of nonpolar atoms.

(2) Contrary to common impression, the backbone atoms of the loop regions are more buried than those of the stems.

(3) The modified groups on the bases increase the exposure areas by as much as 20 per cent of the total base exposure in the molecule, suggesting that such modification may substantially increase the specificity or strength of binding during tRNA-protein recognition. This notion is consistent with the observation that none of the modifications appear to be essential for maintaining the three-dimensional structure.

Flexibility of the molecule

Using a least-squares procedure, Sussman *et al.* (1978) have refined the thermal parameters of bases, riboses and phosphate groups for each residue. The results of this refinement are shown in figure 5.8 where the radius of each circle is proportional to the corresponding group thermal parameters.

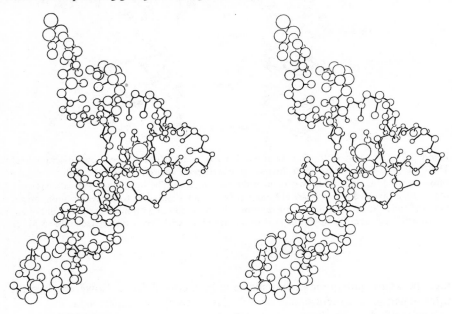

Figure 5.8   Average thermal motions of phosphate, base, and ribose for each residue of yeast tRNA[Phe] shown by circles proportional to the isotropic temperature factors. The larger the circle the higher the thermal vibration of that moiety in the crystal. The thermal motion pattern reflects the flexibilities of different parts of the molecule, the anticodon arm and the amino-acid stem being more flexible than the rest of the molecule.

There are two interesting trends noticeable in the thermal motion of this molecule. The first is that both extreme ends (and three protruding residues, 16, 17, and 47) of the L have very high thermal motion. Average mean square displacements at the extreme ends are about 5.5 Å^2 compared to about 0.8 Å^2 at the corner of the L. This general tendency was also observed in the monoclinic crystal form (Jack *et al.*, 1976). The second is that the average thermal parameter for the bases is smaller than that for the riboses, which in turn is smaller than that for the phosphates. This

Figure 5.9   Three stereoviews of the complete atomic positions in the crystal structure of yeast tRNA[Phe]. Try to follow the progression from 5' end to 3' end by staring for a few minutes, using a pair of stereoglasses, for the top and the bottom pairs. The structural details obtainable from these stereoviews are more accurate and better than any other display methods.

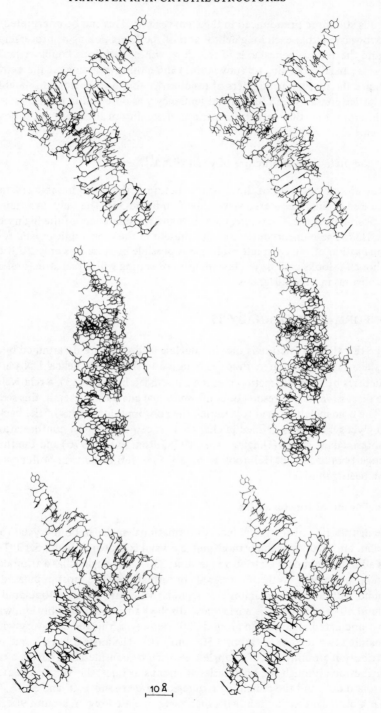

10 Å

trend is still more pronounced in the stem regions. This can be interpreted in one of two ways: either each long helical arm of the L has flexing or processing motion around the average helical axis of each long arm or each long double helical arm of the L is partially opening up (unwinding) and closing (winding) at the extreme end.

Since the thermal parameters of residues are found to be not correlatable with the lattice contacts, and the general tendency was observed in both crystal forms, we interpret that they reveal the intrinsic flexibility of the molecule at both extreme ends of the L.

The complete crystal structure of yeast tRNA^{Phe}

Entire atomic positions of the crystal structure in the orthorhombic crystal can be visualised in the three stereo views shown in figure 5.9. The only uncertainties are the positions of the 3' terminal base A76 and the side chain of the highly modified base H37. These uncertainties are also present in the monoclinic crystal form. Visualisation of the complete molecule is possible because it is only 20 Å thick. Although it looks complex, a few minutes of staring will reveal structural details not obvious in other diagrams.

CONFORMATIONAL ANALYSIS

Since this tRNA structure is the first nucleic acid structure determined by the single crystal x-ray diffraction method (as opposed to the double helical DNA and RNA structures by fibre diffraction methods, discussed in chapter 3), a relatively extensive discussion on the results of conformational analysis is given in this section.

The nomenclature used to describe the conformation angles of the backbone and glycosyl bonds is defined in chapter 1. A complete list of conformation angles at intermediate stages (Quigley *et al.*, 1978; Stout *et al.*, 1976) and the final state (Hingerty *et al.*, 1978; Holbrook *et al.*, 1978) of refinement for both crystal forms have been published.

Distribution of torsion angles

The distribution of the individual conformational angles in both crystal forms are similar, and that from the orthorhombic crystal is shown in figure 5.10. The angle α has a broad, unimodal distribution around 230°. The angle β has a trimodal distribution corresponding to three staggered conformations. The most populated region is around 280° and the remaining two regions, which are sparsely populated, are around 60° and 170°. The angle γ also displays a trimodal distribution, with the most populated region again around 280°. However, this region is very wide. The remaining two regions are around 50° and 170°. The angle δ has a broad unimodal distribution around 170°. The angle ϵ also can be considered as having a trimodal distribution although the boundaries are not as distinct; the most populated region is around 60°, and the remaining two regions are around 170° and 310°. The angle ζ has a bimodal distribution, with the most populated region around 90°, corres-

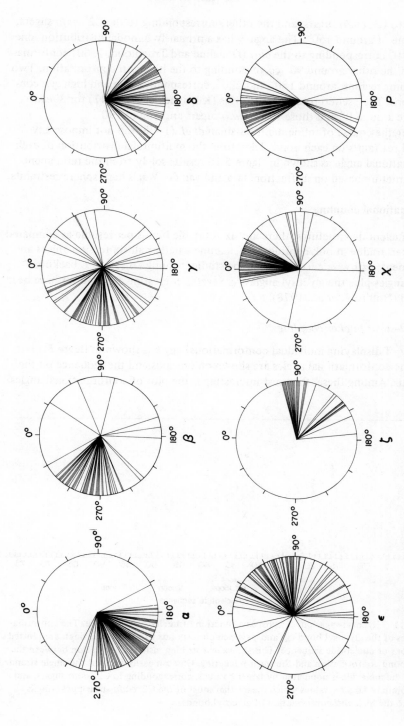

Figure 5.10 The distribution of conformation angles in the yeast tRNA^Phe crystal structure. The conformation angle P represents the pseudo-rotation angle as defined by Kilpatrick *et al.* (1947). The pseudo-rotation angle near 0° represents C3' *endo* riboses, and those around 180° represent C2' *endo* riboses.

ponding to C3′ *endo* sugars, and the other, corresponding to the C2′ *endo* sugars, has a value of around 150°. The angle χ has a primarily bimodal distribution, one around 10° corresponding to the *anti* (Donahue and Trueblood, 1960) conformation, and the other around 90° corresponding to the high *anti* conformation. Two other minor regions, around 180° and 280°, correspond to *syn* and high *syn* conformations. The pseudo-rotation (P) angles (Kilpatrick *et al.*, 1947) for 3′ *endo* sugars are around 0° and those for 2′ *endo* sugars are around 190°.

During the course of refinement, Sussman *et al.* (1978) did not impose any torsional restraints on each bond. Therefore, the multimodal distribution of each conformational angle as shown in figure 5.10 results solely from the refinement of the structure based on diffraction data and van der Waals hard sphere restraints.

Conformational coupling

In polynucleotide structures, there are six rotatable bonds per residue as compared to two per residue in polypeptides. To examine whether all six torsion angles are independent, pair-wise correlation can be studied among 76 sets of 6 backbone torsion angles plus the glycosyl angles, χ. Several interesting correlations have been observed (Holbrook *et al.*, 1978).

Glycosyl–sugar pucker coupling

One way of displaying individual conformational angle is shown in figure 5.11 where the conformational angles are shown on one axis and the sequence on the other axis. Among these, the most interesting is the plot of conformational angles

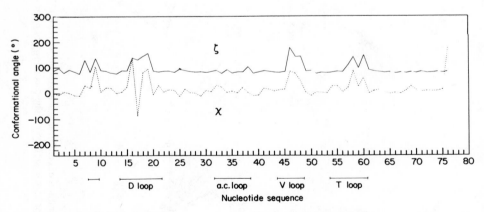

Figure 5.11 Glycosyl-sugar pucker plot as a function of nucleotide sequence. The conformation angles of the glycosyl bond (χ) and ζ (which corresponds to sugar puckering), are plotted as functions of nucleotide sequence. This shows in a striking fashion the relation between the glycosyl bond conformation and the sugar puckering. One can easily see that the single stranded region of the molecule is populated by larger ζ values, corresponding to C2′ *endo* sugars, and the concomitant larger χ values which means that most of the C2′ *endo* sugar puckering is coupled to the high *anti* conformation of glycosyl bonds.

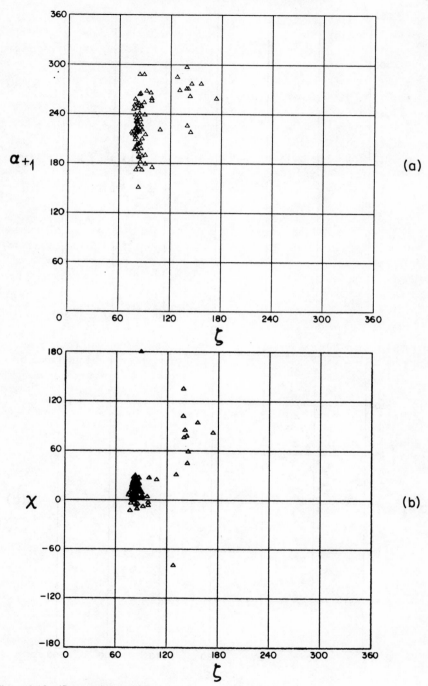

Figure 5.12 (See caption p. 105).

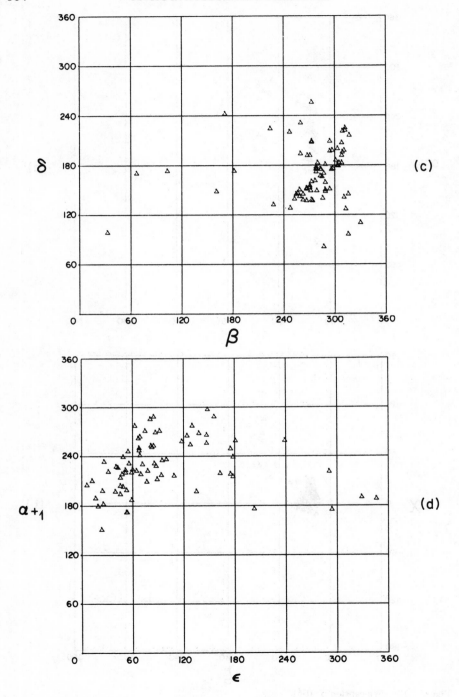

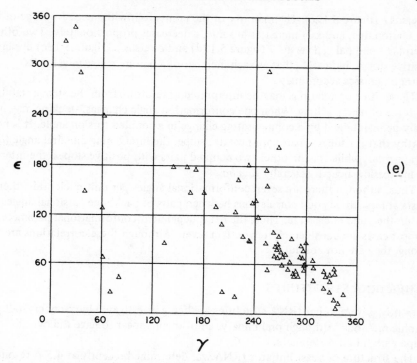

Figure 5.12   (a)–(e) *Intra*nucleotide binary conformational correlation plots with some discernible correlation between a pair of angles are shown. All the angular pairs are from the same nucleotide.

ζ and χ. Two features aré noteworthy: the value of ζ, which distinguishes two possible sugar puckerings, is quite variable at the loops but rather constant in the stems; this variation of ζ is strikingly parallel to the glycosyl bond conformation, χ, except at the 3' end where the refinement is poor.

The α–ζ *plot* shows that C3' *endo* ribose (ζ ~ 85°) can have a fairly wide range of α values while C2' *endo* riboses (ζ ~ 140°) are restricted to a rather narrow range of α values around 270° (figure 5.12a).

The ζ–χ *plot* (figure 5.12b) clearly shows that C3' *endo* riboses are associated with a narrow range (χ ~ 0°) of *anti* conformation for glycosyl bonds, while the C2' *endo* riboses have a fairly wide range of glycosyl conformation, from about 30° to about 140°, and a corresponding *syn* conformation in one case. This correlation can be appreciated from a different point of view in figure 5.11.

Alternate bond torsion angle coupling

Three pairs of conformation angles, each separated by exactly one intervening bond, display a high degree of correlation. The most striking of these is seen in the plot of

γ versus ϵ (figure 5.12c). The majority of the points follow the negative slope of the correlation line, and as γ increases in value, ϵ decreases proportionately. Two other alternate bond pairs, β versus δ (figure 5.12d) and ϵ versus α (figure 5.12e) display positive slopes, indicating that as each member of a pair increases in value its partner increases accordingly.

These three correlations may be interpreted as resulting from the strong stabilising force of base stacking. When one conformation angle changes, stacking may most easily be maintained by a compensating change in a coupled torsion angle. It is noteworthy that for the γ–ϵ pair with negative slope, the intervening dihedral angle (δ) is always *trans*, while for the other two coupled pairs with positive slopes, the intervening torsion angle is generally *gauche*.

Thus, although there are seven conformational angles per unit nucleotide, there exists at least three good correlations between pairs of backbone torsional angles within the same nucleotide. Thus, the net independent conformational parameters in most cases are considerably fewer than seven. Although these correlations are strong, they are not absolute.

OTHER tRNA STRUCTURES

Currently, several other tRNA structures are being investigated by x-ray crystallographic methods. Although preliminary, all of them appear to have the overall shape with minor differences.

The structure of yeast initiator tRNA was determined recently at 4.5 Å resolution by a combination of the multiple isomorphous replacement method, a direct method, and a reciprocal space refinement method (Schevitz *et al.*, 1979). The crystals of this tRNA were grown in a very high salt (2 M ammonium sulphate) solution without any organic solvent as a precipitant. Despite this difference and the fact that this is not an elongation tRNA, the overall structure appears to be the same as that of yeast tRNA^{Phe}. The minor differences as observed at the present time (4.0 Å resolution, R = 25 per cent, J. Sussman and P. Sigler, personal communication) are: the orientation of the anticodon loop with respect to the rest of the molecule is slightly different; the D loop is slightly different due to one additional nucleotide in this loop; no tightly bound magnesium ions are found in this tRNA.

The *E. coli* initiator tRNA structure has been determined by a rotation–translation search method using the crystal structure of yeast tRNA^{Phe} as a searching molecule, and subsequent refinement in real space (N. Woo and A. Rich, personal communication). The preliminary results at 3.5 Å resolution with R = 35 per cent suggest that the overall structure is very similar to that of yeast tRNA^{Phe} with the following minor differences: the anticodon loop conformation, especially around the conserved U residue; the D loop conformation and the interaction between the D and T loop presumably due to the difference in the number of nucleotides in the D loop; some irregularity near the AA stem end.

The *E. coli* tRNA^{Arg} structure has been determined by a rotation–translation search using crystal structure of yeast tRNA^{Phe} as a searching molecule. A preliminary result at 4.5 Å resolution suggests that there may be some difference at the junction between the AC stem and the D stem, but further refinement is needed to ascertain the differences with certainty (Bott and Sundaralingam, personal communication).

The yeast tRNA^{Gly} structure has been determined by the multiple isomorphous replacement method at about 4 Å resolution. The electron density map fitting shows that the overall molecular envelope looks similar to that of yeast tRNA^{Phe} structure, but the base pairs in the amino acid stem are mostly broken, presumably because the crystals are in the presence of about 50 per cent (W/W) dioxane solution. The electron density map revealed that there are four nucleotides in the variable loop rather than three as the nucleotide sequence indicated (Wright, personal communication).

Studies on yeast tRNA^{Asp} structure are in progress by the Strasbourg group, but it is premature at the present time to make any comparisons.

GENERALITY OF THE YEAST tRNA^{Phe} STRUCTURE

Conservation of tertiary structure of tRNA

The first indication that all tRNAs have the same overall structure comes from the observation that all tRNA sequences can be arranged into a cloverleaf model. The second convincing indication was derived from the crystal structure of yeast tRNA^{Phe}, where *most* of the conserved and semiconserved bases are found to be involved in tertiary hydrogen bonds. Moreover, most of these tertiary base pairs in yeast tRNA^{Phe} can be replaced with equivalent tertiary base pairs for other tRNAs without disturbing the backbone structures significantly, lending strong credence to the notion that the overall structure of all tRNAs may be the same (Kim *et al.*, 1974; Klug *et al.*, 1974). A list of such replacable tertiary base pairs, based on model building, is given in table 5.6. Thus, the crystal structure of yeast tRNA^{Phe} can be considered as a representative structure of all tRNAs.

(G, C)-rich helical stems

The (G, C)-contents in the stems of tRNAs are considerably higher than 50 per cent. Furthermore, one end of each stem is always very high in (G, C)-content except the anticodon stem, where a penultimate base pair is most often a G·C pair (figure 5.13). They are G1 C72, G20 C25, G53 C61 and G30 C40 in yeast tRNA^{Phe} At equivalent positions in all other tRNAs, the first three are G·C pairs in most of the cases, but the fourth one is either a G·C or C·G pair. The frequencies of occurrence of bases in tRNA sequence are listed in table 5.7.

The (G, C)-content of the long helix formed by the acceptor and T stems is highest at the ends and averages 55 per cent in the middle. Such richness in G-C pairs at extremes may stabilise the long helix from unwinding at the ends (fraying).

The lowest (G, C) frequency (51 per cent) among all stems is found at the base

Table 5.6   Replaceable tertiary base pairs in $D_4 V_5$ tRNAs

In yeast tRNA^{Phe} residue number (bases)	In all $D_4 V_5$ tRNAs (51 sequences examined)
8·14(U·A)	51 U·A
9·23(A·A)	33 A·A
	10 G·C
	3 G·G
	3 A·C[b]
	1 A·G
	1 G·A
10·45(G·C)	39 G·G
	2 G·U
	1 G·A[a]
15·48(G·C)	42 G·C
	7 A·U
	1 G·A[b]
	1 A·C
18·55(G·ψ)	51 G·ψ
19·56(G·C)	51 G·C
22·46(G·G)	46 G·G
	3 A·A
	1 C·A[b]
	1 G·U
26·44(G·A)	29 G·A
	13 A·G
	3 A·C
	2 A·U
	1 C·U
	1 G·G[b]
	1 A·A
	1 ψ·G
54·58(T·A)	51 T·A

[a]Cannot make approximately equivalent hydrogen bonds.
[b]Require considerable twisting around the hydrogen bond.

pair immediately next to the anticodon loop, suggesting that this base pair may be easily broken during tRNA interaction with the ribosome to provide additional flexibility to the anticodon loop.

'Conserved presence' and 'conserved absence' of bases in tRNA

As mentioned earlier, certain bases are conserved at particular locations in nucleotide sequence and most of them are involved in forming hydrogen bonds to maintain the tertiary structure of tRNA. The converse question may be asked: are certain bases always *absent* at particular locations in nucleotide sequence and if there are, what is their functional significance? Figure 5.14 shows bases and locations of such 'conserved presences' and 'conserved absences'.

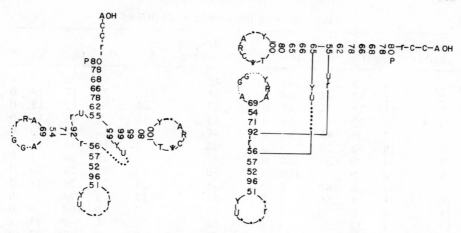

Figure 5.13 G·C base pair frequencies in per cent occurring at each position of the stems for all transfer RNA so far sequenced.

Among the conserved bases, U8, A14, G18, G19, A21, T44, ψ55, C56, A58 form tertiary hydrogen bonds, but A16, U33, C74, C75, A76 are not involved in tertiary base pairing. The latter three are essential for aminoacylation. The first two, A16 and U33, probably are essential for ribosomal function(s), which should be common to all tRNAs.

Among the conserved absence of certain bases, the absence of Y9, A12 and U23 can be explained from the necessity of forming a replaceable tertiary base triple 9-23-12. The same reasoning can be applied to the absence of Y15 and R48. The conserved absence of C59 and R60 may be due to the tight spatial requirement, that is, the residues 59 and 60 are buried in a pocket inside the tertiary structure of tRNA, forming two hydrogen bonds by phosphate 60 with the base 61 and ribose 58. The remaining conserved absences must have other functional reasons. For example, the absence of A34 is due to the conversion of adenine to inosine, presumably because the amino group of adenine is incompatible with necessary interactions with, say, the ribosome. In general, reasons for many of the conserved absences are not understood at the present time.

CONCLUDING REMARKS

The crystal structure of yeast tRNA^{Phe} serves two major, useful purposes: one to help understand structure-function relationship of tRNA, and the other, as a model for understanding the three-dimensional structures of nucleic acids in general. The structure revealed an architectural design by which a relatively short nucleic acid can be folded up to form a compact, stable structural entity that has many diverse functions in living cells. The structure also has a functional design in that three important functions in ribosomes are associated with three well-separated regions of the structure.

TOPICS IN NUCLEIC ACID STRUCTURE

Table 5.7 Base frequencies in all tRNAs[a]

Residue number	A	C	G	U	N	Residue number	A	C	G	U	N
1	6	9	74	11	0	40	2	69	26	3	0
2	5	37	44	14	0	41	19	35	17	29	0
3	11	38	32	19	0	42	23	6	52	18	0
4	7	22	45	26	0	43	32	7	46	15	0
5	17	29	43	12	0	44	50	12	18	20	0
6	18	30	29	23	0	48	2	80	1	17	0
7	24	0	55	20	0	49	24	26	46	4	0
8	0	0	0	100	0	50	12	38	30	20	0
9	52	6	40	1	0	51	23	24	39	13	0
10	1	1	96	2	0	52	17	3	77	3	0
11	5	69	0	26	0	53	0	0	100	0	0
12	4	27	27	43	0	54	6	0	0	93	1
13	10	60	13	18	0	55	0	0	3	97	0
14	100	0	0	0	0	56	0	100	0	0	0
15	21	0	79	0	0	57	22	0	78	0	0
21	97	0	2	1	0	58	100	0	0	0	0
22	28	1	64	7	0	59	41	5	23	30	0
23	44	27	26	4	0	60	7	20	0	72	0
24	21	0	73	5	0	61	0	100	0	0	0
25	0	86	1	13	0	62	3	77	3	17	0
26	39	4	55	1	0	63	10	39	28	23	0
27	10	45	13	33	0	64	17	22	41	19	0
28	18	50	6	24	1	65	3	32	27	38	0
29	29	16	35	20	0	66	19	54	1	26	0
30	0	24	72	3	0	67	14	28	37	21	0
31	43	32	19	6	0	68	7	39	34	19	0
32	0	68	0	32	0	69	20	37	28	15	0
33	0	1	0	99	0	70	19	27	38	16	0
34	0	23	51	22	3	71	13	37	38	12	0
35	41	21	16	21	0	72	15	74	3	7	0
36	28	30	13	30	0	73	65	1	23	11	0
37	79	3	18	0	0	74	0	100	0	0	0
38	64	19	3	14	0	75	0	100	0	0	0
39	4	19	32	45	0	76	100	0	0	0	0
						Total	22	29	28	20	0

[a]The numbering system is such that the nucleotides in stem and conserved positions of all tRNAs have the same numbers. For example, the anticodon triplets have the numbers 34, 35 and 36; the 3′ terminal A is 76. N is for unknown bases.

The crystal structure determined by x-ray crystallographic methods is necessarily static and no knowledge of the dynamic aspect of the molecule can be obtained *directly* either for a free molecule or for the molecule interacting with other macromolecules. Nevertheless, the crystal structure provides a solid and dependable starting point for studies on tRNA function at all levels.

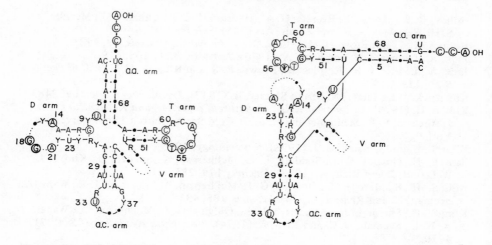

Figure 5.14 'Conserved presence' and 'conserved absence' of bases in $D_4 V_5$ class tRNAs (tRNAs with four base pairs in the D stem and five bases in the V loop). When a particular base is found at a certain position in more than 90 per cent of the tRNA sequences (conserved presence), it is shown as a circle. If a base occurs at a particular position less than 10 per cent of the time (conserved absence), it is shown without a circle.

ACKNOWLEDGEMENTS

Work performed in the author's laboratory has been supported by the grants from National Institutes of Health (CA-27454, K04-CA-00352) and the National Science Foundation (PCM76-04248). Thanks are extended to Joel Sussman, Steve Holbrook, Wade Warrant, George Church and Charles Alden for their critical contributions in various stages of crystallographic refinement of yeast tRNA[Phe] structure and interpretation of the results, some of which are described here, and Beth Klingel for the preparation of the manuscript.

REFERENCES

Alden, C. and Kim, S. H. (1979). *J. molec. Biol.,* **131**, 411.
Clark, B. F. C. (1977). *Prog. Nucl. Acid Res. Mol. Biol.,* **20**, 1.
Donahue, J. and Trublood, K. N. (1960). *J. molec. Biol.,* **2**, 363.
Erdmann, V. A. (1978). *Prog. Nucl. Acid. Res. Mol. Biol.,* **18**, 45.
Fresco, J. R., Adams, A., Ascione, R., Henley, D. and Lindhal, T. (1966). *Cold Spring Harb. Symp. quant. Biol.,* **31**, 527.
Haselkorn, R. and Rothmann-Denes, L. R. (1973). *A. Rev. Biochem.,* **42**, 397.
Hingerty, B., Brown, R. S. and Jack, A. (1978). *J. molec. Biol.,* **124**, 523.
Holbrook, S. R., Sussman, J. L., Warrant, W. R., Church, G. M. and Kim, S. H. (1977). *Nucl. Acids Res.,* **4**, 2811.
Holbrook, S. R., Sussman, J. L., Warrant, R. W. and Kim, S.-H. (1978). *J. molec. Biol.,* **123**, 631.

Holley, R. W., Apgar, J., Everett, G. A., Madison, J. T., Marquisee, M., Merrill, S. H., Penswick, J. R. and Zamir, A. (1965). *Science,* **147**, 1462.

Holmes, K. C. and Blow, D. M. (1965). *Meth. biochem. Analysis,* **13**, 113.

Jack, A., Ladner, J. E. and Klug, A. (1976). *J. molec. Biol.,* **108**, 619.

Jack, A., Ladner, J. E., Rhodes, D., Brown, R. S. and Klug, A. (1977). *J. molec. Biol.,* **111**, 315.

Kilpatrick, J. E., Pitzer, K. S. and Spitzer, R. (1947). *J. Am. chem. Soc.,* **69**, 2483.

Kim, S.-H. (1979). In *Transfer RNA: Structure, Properties, and Recognition* (eds. J. Abelson, P. R. Schimmel and D. Söll), Cold Spring Harbor Laboratory, New York.

Kim, S. -H. and Quigley, G. J. (1978). *Enzymology,* **59**, 3.

Kim, S. -H., Quigley, G. J., Suddath, F. L., McPherson, A., Sneden, D., Kim, J. J., Weinzierl, J. and Rich, A. (1973). *Science,* **179**, 285.

Kim, S. -H., Suddath, F. L., Quigley, G. J., McPherson, A., Sussman, J. L., Wang, A., Seeman, N. and Rich, A. (1974). *Science,* **185**, 435.

Kim, S. -H., Sussman, J. L., Suddath, F. L., Quigley, G. J., McPherson, A., Wang, A. H. L., Seeman, N. C. and Rich, A. (1974). *Proc. natn. Acad. Sci. U.S.A.,* **71**, 4970.

Klug, A., Ladner, J. and Robertus, J. D. (1974). *J. molec. Biol.,* **89**, 511.

LaRossa, R. and Söll, D. (1978). In *Transfer RNA* (ed. S. Altman), M.I.T. Press, Cambridge, Massachusetts, p. 136.

Lee, B. K. and Richards, F. M. (1971). *J. molec. Biol.,* **55**, 379.

Nishimura, S. (1978). In *Transfer RNA* (ed. S. Altman), M.I.T. Press, Cambridge, Massachusetts, p. 168.

Quigley, G. J., Teeter, M. M. and Rich, A. (1978). *Proc. natn. Acad. Sci. U.S.A.,* **74**, 64.

Rich, A. and RajBhandary, U. L. (1976). *A. Rev. Biochem.,* **45**, 805.

Schevitz, R. W., Podjarny, A. D., Krishnamachari, N., Hughes, J. J., Sigler, P. B. and Sussman, J. L. (1979). *Nature, Lond.,* **278**, 188.

Sprinzl, M., Grüter, F. and Gauss, D. H. (1978). *Nucl. Acids Res.,* suppl., r15.

Stein, A. and Crothers, D. (1976). *Biochemistry,* **15**, 157.

Stout, C. D., Mizuno, H., Rao, S. T., Swaminathan, P., Rubin, J., Brennan, T. and Sundaralingam, M. (1978). *Acta crystallogr.,* **B34**.

Stout, C. D., Mizuno, H., Rubin, J., Brennan, T., Rao, S. T. and Sundaralingam, M. (1976). *Nucl. Acids Res.,* **3**, 1111.

Suddath, F. L., Quigley, G. J., McPherson, A., Sneden, D., Kim, J. J., Kim, S. H. and Rich, A. (1974). *Nature, Lond.,* **248**, 20.

Sussman, J. L., Holbrook, S. R., Warrant, R. W., Church, G. M. and Kim, S. H. (1978). *J. molec. Biol.,* **123**, 607.

6

The solution structure of transfer RNA studied by proton NMR

Brian R. Reid

INTRODUCTION

The adaptor role of transfer RNA in protein biosynthesis was discovered some twenty years ago (Hoagland *et al.*, 1957). Fifteen years ago the nucleotide sequence of the first tRNA molecule was determined by Holley and coworkers (Holley *et al.*, 1965). The primary sequence led to the prediction of the well-known cloverleaf secondary structure which has been amply corroborated by the determination of over 80 tRNA sequences in the last ten years (Sprinzl *et al.*, 1978). Elucidation of higher order folding into a defined three-dimensional structure was complicated by the difficulty in obtaining well-ordered crystals of this highly charged polymer. Six years ago Kim and Rich and coworkers at M.I.T. (Kim *et al.*, 1974) and the Klug group at M.R.C. Cambridge (Robertus *et al.*, 1974) reported the tertiary interactions in the crystal structure of yeast tRNA^{Phe} at intermediate crystallographic resolution. The high-resolution refined structure has been reported by four laboratories (Jack *et al.*, 1976; Quigley *et al.*, 1975; Stout *et al.*, 1978; Sussman *et al.*, 1978); the three-dimensional folding and tertiary interactions are in virtually complete agreement in these four refined structures and are discussed in detail in chapter 5.

Despite the existence of a detailed three-dimensional structure for yeast tRNA^{Phe}, and a large body of kinetic information on the aminoacylation of this molecule by its cognate aminoacyl-tRNA synthetase, there is still no good explanation of *why* the genetic code is as it is known to be, that is, the aspects of this particular polynucleotide that make it uniquely suited to be the adaptor for the amino acid phenylalanine are not understood. This latter question obviously lies at the heart of *how* the genetic code works. Although there are no good theories to explain this process it is apparent that the answer will come only from studies of the conformation and dynamics of this molecule in solution, that is, under conditions in which it carries out its biological adaptor function. The tools for such studies are obviously spectroscopic ones. During the last ten years a variety of spectroscopic

methods have been applied to the problem of tRNA solution structure. The
majority of these have been optical studies of conformational changes utilising
ultraviolet hyperchromicity which suffer from poor spectroscopic resolution,
resulting in relatively low-resolution information; the results from such studies have
been reviewed by Crothers and Cole (1978). A more detailed analysis of tRNA
solution conformation requires resolution of the various spectroscopic signals and
by far the most promising technique in this direction is nuclear magnetic resonance.

The average tRNA molecule contains almost 1000 hydrogen atoms which
immediately poses resolution problems when the limited range of chemical shifts
in proton NMR is considered. There are several hundred ribose protons which lie in
a poorly resolved group quite close to the water resonance around −5 ppm (from
the reference DSS resonance). Further downfield in the −6 ppm to −8.5 ppm
region the aromatic CH protons and exocyclic amino protons of the heterocyclic
bases are found. The latter are exchangeable and can be eliminated by working in
D_2O solvents. However, in a typical tRNA the remaining aromatic resonances
number approximately one hundred protons which are poorly resolved in this
crowded spectral region since they have relatively broad line-widths due to the long
rotational correlation times of the native tRNA polymer. The aromatic NMR spectrum of tRNA has been studied by MacDonald et al. (1964) and by Smith, Shulman
and coworkers (Smith et al., 1968, 1969) but little fine detail has emerged from
such studies. The most informative region of the NMR spectrum of tRNA is the
extreme low field region between −11 ppm and −15 ppm; the next most useful
spectral region is the extreme high field spectrum from 0 ppm to −4 ppm containing the aliphatic methyl and methylene resonances of modified nucleotides.

LOW FIELD HYDROGEN BOND NMR

In the NMR spectra of mononucleotides the most deshielded proton is the imino
hydrogen attached directly to the heterocyclic ring nitrogen. The ring NH proton
exchanges quite rapidly with water and hence can only be observed in monomers in
aprotic solvents. The predominant tautomers of cytidine and adenosine do not
contain a ring NH proton, in contrast to uridine (U N3H) and guanosine (G N1H).
These protons resonate close to −11 ppm in dry DMSO. Hydrogen bonding of U to
the complementary A, and of G to the complementary C, causes further deshielding to
even lower field (Katz and Penman, 1966). As shown in figure 6.1 each complementary
base pair contains only one hydrogen-bonded ring NH proton so that the low field
spectrum intensity (−11 ppm to −15 ppm) of a polynucleotide should immediately
reflect the number of base pairs provided the base-pair helix lifetime is sufficiently
long. The chemical shift differences between helix and coil states are approximately
1−2 ppm and a given ring NH proton must remain in the helical base-paired state
for approximately 5 ms or longer to generate a discrete resonance characteristic of
that state (the resonance of the exposed coil state effectively time-averages into the
solvent peak in aqueous solution via rapid proton exchange). A further point worth
noting in figure 6.1 is that U N3H resonates around 0.7 ppm to lower field than

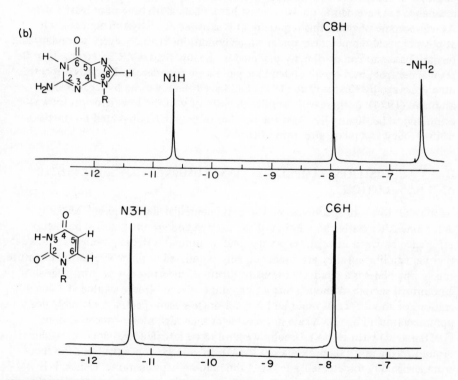

Figure 6.1   (a) The hydrogen bonding of complementary base pairs showing the ring NH of G and the ring NH of U in bold face. (b) The low field portion of the NMR spectrum of guanosine and uridine in dry d_6 DMSO showing the extremely deshielded chemical shift of the ring NH proton.

G N1H. This chemical shift difference is further increased by complementary base pair formation in which U N3H is strongly downfield-shifted by the in-plane large ring current of adenine whereas G N1H is only mildly shifted by the weak ring current of cytosine.

Because of dynamic helix-coil effects it is apparent that hydrogen-bonded ring NH protons in the 'inaccessible' helical state will still exchange with solvent in aqueous solution. Thus the extreme low field proton spectrum cannot be observed in D_2O solutions due to exchange of these protons for deuterons during the time of sample preparation, spectrometer tuning, and so on. Such hydrogen bonds must be studied in H_2O solvents which creates special problems due to the enormous (110 M) intensity of the solvent proton resonance. The advantages of being able to directly study the Watson-Crick hydrogen bonds themselves far outweigh the disadvantages since there are several methods of overcoming the large solvent proton peak. As mentioned earlier, in order to generate a hydrogen-bonded proton resonance the base pair in question must have a helical lifetime of at least 5 msec. As will become obvious, this requirement is satisfied for tRNA in aqueous solution at physiological temperature, under which conditions virtually every secondary and tertiary base pair can be directly detected in the low field NMR spectrum. Low field resonances from hydrogen-bonded base pairs were first detected in tRNA spectra nine years ago by Kearns *et al.* (1971) and have been reviewed by Kearns and Shulman (1974); in these early studies the intensity of the low field spectra was estimated to be slightly less than the number of base pairs predicted on the basis of the Holley secondary structure of tRNA.

DATA ACQUISITION, SAMPLE SIZE, INSTRUMENTATION, LINEWIDTHS AND RESOLUTION

As in other forms of spectroscopy, the Boltzmann distribution $\exp(-\Delta E/kT)$ determines the relative population of the excited and ground states. The energy difference between these states is very small in normal NMR experiments so that at thermal equilibrium they are almost equally populated at physiological temperature; this is why NMR is a relatively insensitive form of spectroscopy requiring large amounts of sample. Although many small molecules are freely soluble at concentrations of about 50 mM required for good single-sweep spectra, most biological polymers (including tRNA) are not soluble at such high molar concentrations (1250 mg ml^{-1} for tRNA). Usually we have to be satisfied with concentrations around 25 mg ml^{-1} (approximately 1 mM) and at such concentrations the spectrum is virtually undetectable amongst the random baseline noise. Hence, it is necessary to superimpose several hundred spectra which requires a spectrometer with a signal-averaging computer of average transients (CAT). The actual sensing probe of a normal proton spectrometer is focussed on a volume of 5 mm diameter by approximately 10 mm height so that 1 mM samples in a volume of 0.2 ml (that is, 5 mg of tRNA) can be signal-averaged to a good signal-to-noise ratio in a reasonable period of time.

In a classical frequency sweep spectrum the problems of the large solvent peak can be avoided by simply not sweeping through this region of the spectrum. This is of no use for resonances close to the water frequency but in low field tRNA spectroscopy we can set the upper limits of the frequency sweep at approximately −9 ppm and repetitively sweep only the −15 ppm to −9 ppm region. Excessively rapid sweep rates lead to peak distortion, especially for narrow lines. The linewidth $(1/\pi T_2)$ of a resonance is determined by proton relaxation processes and in tRNA low field spectra the two most important are nitrogen quadrupolar relaxation and the long rotational correlation time of the polymer (slow tumbling); these processes rapidly relax protons leading to relatively broad tRNA low field lines of around 25–30 Hz at physiological temperature. Therefore sweep rates of 200–300 Hz s^{-1} can be used without serious peak distortion. This leads to dwell times of around 0.1 s on each resonance and the amplitude of the radio frequency must be reduced to avoid saturating the resonances during these relatively long dwell times; this restriction further reduces sensitivity. Hence a sweep of 2500 Hz might take 10 s and signal-averaging 700 such sweeps will take about two hours using classical CW (continuous wave) spectroscopy. However, this time period can be reduced to 10 minutes or less by rapid-sweep correlation spectroscopy in which the peak distortion is corrected by cross-correlation in a computer (Dadok and Sprecher, 1974). Even greater sensitivity can be achieved in theory by using the Fourier advantage in pulsed FT (Fourier transform) spectroscopy. FTNMR of low concentrations of polymers in water has been a difficult or impossible task due to computer dynamic range problems associated with large solute-solvent concentration differences (five orders of magnitude) especially in the case of exchangeable protons in equilibrium with water protons. However, Redfield has recently introduced pulsed FTNMR methods which greatly reduce the intensity of the solvent peak without perturbing the water protons; this innovative technique has opened up the field of time-resolved NMR spectroscopy and will be discussed later in connection with helix dynamics.

A further point to be mentioned is the effect of magnetic field strength in resolving spectra of polymers such as tRNA. As mentioned earlier, the low field hydrogen-bonded ring NH resonances in tRNA have inherent linewidths of approximately 30 Hz at physiological temperature. These linewidths remain 30 Hz regardless of magnetic field strength and the majority of tRNA low field resonances are concentrated between −14 ppm and −12 ppm. At magnetic field strengths of 63.4 kilogauss (270 MHz proton frequency) this 2 ppm window would only encompass 540 Hz and it would be difficult to resolve 25 lines of 30 Hz. At 400 MHz this same 2 ppm region of the spectrum would encompass 800 Hz and at 600 MHz (141 kilogauss) it would represent a 1200 Hz range in which 30 Hz lines separated by only 0.05 ppm would be partially resolved. Hence, it is important, especially for relatively broad line biological polymer spectra, to work at the highest available magnetic field strength in order to effect maximal resolution. The increased resolution in the low field proton spectrum of *E. coli* tRNA$_1^{Val}$ upon going from 360 MHz to 600 MHz is shown in figure 6.2; these spectra should be compared with the spectra of the same molecule at 300 MHz (Bolton and Kearns, 1976) and at 270 MHz (Reid *et al.*, 1975). In addition to the increased resolution an added benefit from working at high magnetic field strength is the increase in sensitivity due to the greater energy separation between ground and excited states which results in shorter signal-averaging times to attain a given signal-to-noise ratio.

INTEGRATION OF SPECTRA

The number of base pairs with helical lifetimes of at least 5 msec can be directly
determined from the intensity of the hydrogen bonded ring NH low field spectrum
between −11 ppm and −15 ppm. Resonances between −9 ppm and −11 ppm are
derived from various sources including GU 'wobble' base pairs, ring NH protons
hydrogen bonded to anomalous acceptors such as phosphates, and extremely deshielded aromatic CH protons.

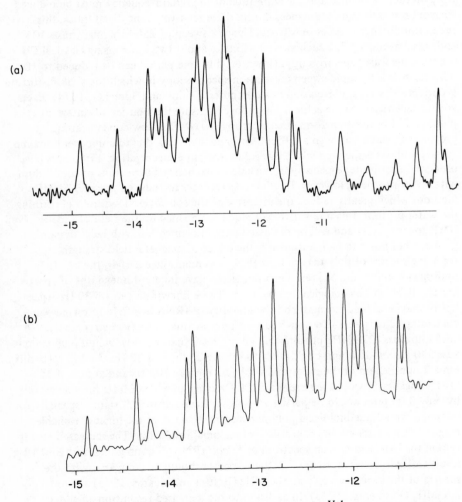

Figure 6.2   The low field NMR spectrum of 6 mg of *E. coli* tRNA$_1^{Val}$ in 20 mM sodium phosphate, 100 mM NaCl pH 6.8 at (a) 360 MHz and 37° and (b) 600 MHz and 22°. Both were
signal-averaged under fast-sweep correlation spectroscopy conditions. (The 600 MHz spectrum
was taken by J. Dadok and A. Bothner-By on a sample only 50 per cent thiolated at residue 8;
note the half proton peaks at −14.9 ppm and −14.25 ppm.)

In spectra which are inherently well resolved or have been resolved by very high field strength, the total low field intensity is fairly obvious. For instance, the *E. coli* tRNA$_1^{Val}$ spectrum reveals 27 or 28 base pairs (figure 6.2). For less resolved spectra the intensity can be determined by computer simulation. In this technique a series of unit intensity peaks are given the experimental linewidth and Lorentzian line-shape and are then reiteratively moved in the x axis (chemical shift) until they duplicate the experimental spectrum. In this way the relative intensities of shoulders and complex peaks can be evaluated. An example is shown in figure 6.3; the com-

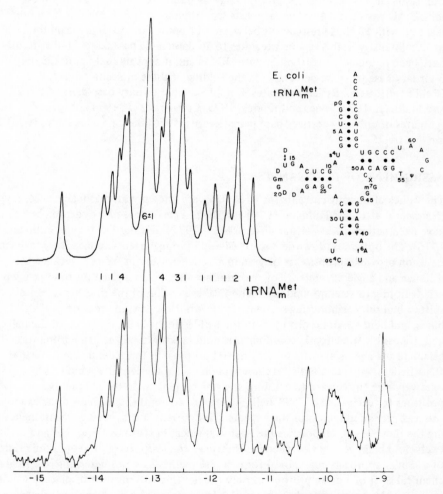

Figure 6.3   The 360 MHz low field NMR spectrum of *E. coli* tRNA$_m^{Met}$ in the presence of excess magnesium. The upper spectrum is a computer simulation of 27 Lorentzian lines of the experimental height and linewidth, the intensities of the various peaks are designated above the experimental spectrum.

puter simulation and intensity analysis of the experimental *E. coli* tRNA$_m^{Met}$ spectrum indicates 27 stable base pairs detectable in solution. *E. coli* tRNA$_m^{Met}$ and tRNA$_1^{Val}$ both contain only 20 base pairs in their cloverleaf secondary structure.

The marked differences in chemical shift of the low field resonances between tRNAs of differing sequence is principally caused by nearest neighbour ring current shifts from adjacent stacked bases. Thus tRNA$_1^{Val}$ contains 5 AU and 15 GC pairs and tRNA$_m^{Met}$ contains 6 AU, 13 GC and 1 Aψ pair, yet is less resolved. However, one can find other tRNAs of similar base pair content which, because their different sequence leads to fortunate neighbouring ring current shifts, are inherently well resolved. For example, *E. coli* tRNA$_3^{Ala}$ is similar to tRNA$_m^{Met}$ in that it contains 6 AU pairs and 14 GC pairs, yet its spectrum, even at 360 MHz, is far more resolved with 13 single resonances between -11 ppm and -15 ppm; again the proton intensity in this region integrates to 26 detectable base pairs. Similar results have been obtained with 10 other pure tRNAs and it appears that approximately 6 tertiary base pairs are detectable in the folding of these molecules in solution (Reid *et al.*, 1977). Earlier estimates of no detectable tertiary base pairs (Kearns and Shulman, 1974; Jones and Kearns, 1975; Crothers *et al.*, 1974) or revised estimates of only 2–4 tertiary base pairs (Kearns, 1976; Bolton and Kearns, 1976) are incorrect.

ASSIGNMENT OF RESONANCES

The values of the ring currents for RNA bases are known with reasonable accuracy (Geissner-Prettre and Pullman, 1970) and the effects of base-pair hydrogen bonding have been determined (Geissner-Prettre *et al.*, 1977; Geissner-Prettre and Pullman, 1976). The upfield or downfield shifts caused by ring current-induced local magnetic fields on proximal protons are diagrammed schematically in figure 6.4. Since the distance and angle dependence of the ring-current shift on a given proton is known it is tempting to attempt a purely theoretical evaluation of the ring NH chemical shifts when the neighbouring sequence is known. However, the geometry of the bases, and hence θ and r, change with the helical pitch. Thus, the pitch of the helix and, if possible, the atomic coordinates should be known before attempting to calculate the chemical shifts of the ring NH protons. In the general case of regular RNA helices only certain discrete screw pitch values are permitted and it is relatively straightforward to combine the ring currents with the defined base positions for each type of RNA helix. Arter and Schmidt (1976) have done exactly this and presented tables for the expected ring current shifts on all protons, including the ring NH, for all possible base-pair sequences in tenfold, elevenfold and twelvefold helix geometries. Giessner-Prettre *et al.* (1976) have presented the results of a similar approach in graphical form. These methods do not determine the final chemical shift of a given proton but only the net ring current shift displacement from its 'inherent unshifted position'; this latter parameter must be determined independently, usually by trial-and-error fitting to experimental spectra containing reliably assigned lines. A further complication in the case of tRNA is the possibility that some of the base pairs may not be in 'regular' helix geometry.

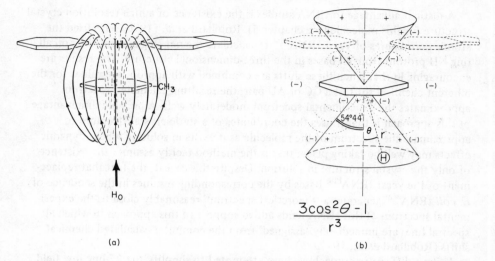

$$\frac{3\cos^2\theta - 1}{r^3}$$

Figure 6.4   (a) Diagrammatic conceptualisation of the induced magnetic field from the ring current of a pyrimidine stacked below an adjacent base pair. (b) Illustration of the distance and angle dependence of the ring current shift for a proton located at the point H.

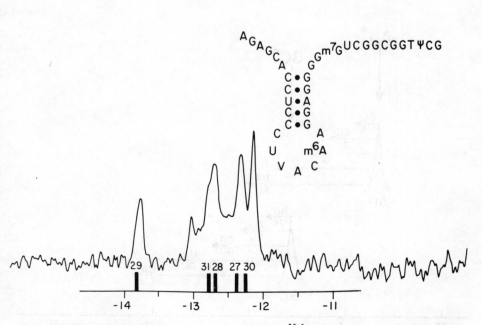

Figure 6.5 The 360 MHz NMR spectrum of *E. coli* tRNA$_1^{Val}$ fragment 21–57 containing the anticodon helix. The numbered bars refer to the chemical shifts of the corresponding base pairs predicted from elevenfold RNA helical geometry as described in the text.

A distinct advantage to tRNA studies is the existence of a high-resolution crystal structure for this molecule (see chapter 5). Robillard *et al.* (1976) have used the atomic coordinates of yeast tRNA[Phe] to calculate the net ring current shift on all ring NH protons from all bases in the three-dimensional structure. The results are encouraging in that when these shifts are combined with appropriate values for the inherent chemical shift of a GC or AU pair, the resulting calculated spectrum approximates the experimental spectrum moderately well. A possible disadvantage of this approach is that it uses the coordinates of a static crystal structure to approximate the spectrum of the molecule as it exists in solution where dynamic effects may well be taking place, that is the method tacitly assumes the existence of only the crystal structure in solution. Despite this caveat, the fact that replacement of the yeast tRNA[Phe] bases by the corresponding residues in the sequence of *E. coli* tRNA$_1^{Val}$ generates a 'theoretical spectrum' reasonably close to the experimental spectrum of this tRNA lends added support to this approach in which all spectral lines are immediately 'assigned' from the computer-calculated chemical shifts (Robillard *et al.*, 1977*a, b*).

Using a different approach we have attempted to simplify the 27-line low field spectra of several tRNA species by chemical dissection and isolation of the short component helices. These short 4–6 base-pair hairpin helices generate simple 4–6 line low field spectra which are usually resolved enough to permit assignments based only on the sequence and a semiquantitative understanding of purine and

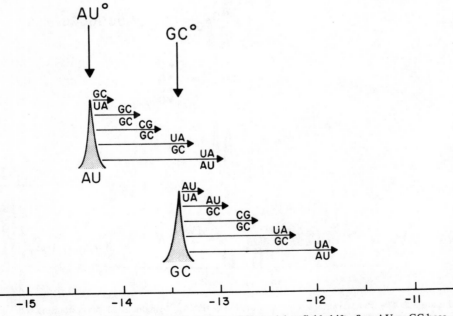

Figure 6.6 Diagrammatic representation of the differential upfield shift of an AU or GC base pair in a variety of sequence environments.

pyrimidine ring currents. The advantage of this approach is that it makes no prior
assumptions about the crystallographic coordinates necessarily pertaining in
solution; the disadvantage is that some helices, when dissected from the intact
tRNA, may relax into a helical structure different from that in the whole molecule.
An example of this approach is given in figure 6.5 which shows the low field spectrum of the *E. coli* tRNA$_1^{Val}$ residue 21–57 fragment containing the 5 base pair
anticodon helix. The fact that all helices studied containing purely GC pairs do not
contain resonances lower than −13.4 ppm immediately assigns the resonance at
−13.8 ppm to the single AU pair between positions 29–41. This is also in agreement with the monomer studies which indicate that AU pairs should resonate at
lower field than GC pairs. In RNA helical geometry A41 (not U29) is directly on
top of the C30–G40 ring NH and, since A has by far the strongest ring current,
CG30 should be the most upfield shifted resonance (at −12.2 ppm). Using such
experimentally determined chemical shifts it is possible to go back and check the
ring current shifts predicted by the various theoretical methods. We have carried
out such analyses on the rT helices and anticodon and DHU helices of *E. coli*
tRNA$_1^{Val}$, tRNA^{Lys} and tRNA^{Phe} (Reid *et al.*, 1979). The presence of purines in
the upper right (3′) and lower left (5′) nearest neighbour positions causes much

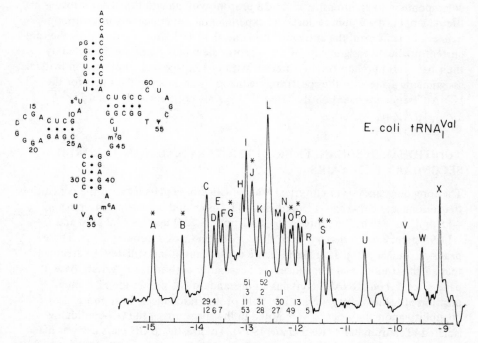

Figure 6.7 The 360 MHz NMR spectrum of *E. coli* tRNA$_1^{Val}$ at 37° in the presence of limiting
amounts of magnesium. The numbers under the peaks refer to the corresponding secondary
base pairs assigned by a combination of hairpin helix fragment spectra and ring current shift
calculations (in the case of the acceptor helix). Additional resonances are denoted by asterisks.

stronger upfield shifts than pyrimidines in these positions as expected for RNA helix geometry. For instance, a GC ring NH in the sequence AU–GC–UA resonates at approximately −13.2 ppm whereas in the sequence UA–GC–AU the GC resonance is found at approximately −11.8 ppm. These sequence effects are diagramatically summarised in figure 6.6. Arter and Schmidt (1976) have pointed out that the second order next-to-nearest neighbour effects, although attenuated by distance to around 10 per cent of the nearest neighbour effects, are not negligible. We have had quite good success in fitting our experimental data by using inherent unshifted AU° and GC° values of −14.35 ppm and −13.45 ppm, respectively, combined with the ring NH shifts predicted for elevenfold RNA geometry published by Arter and Schmidt (1976). These theoretical shifts for the anticodon helix of $E.\ coli$ tRNA$_1^{Val}$ are shown by the black bars on the abscissa in figure 6.5.

The spectrum of $E.\ coli$ tRNA$_1^{Val}$ and other tRNA species is dependent on the level of magnesium ion in the solvent. In the complete absence of magnesium the seven protons between −11.8 ppm and −12.4 ppm are poorly resolved whereas in the presence of excess magnesium the region around −13 ppm is poorly resolved (Reid $et\ al.$, 1979). Figure 6.7 shows the spectrum at 360 MHz in an EDTA buffer containing limiting amounts of magnesium, resulting in maximum overall resolution; there are 21 resolved or partially resolved peaks between −15 ppm and −11 ppm corresponding to an intensity of 27–28 protons with an additional 4–5 protons in the −11 ppm to −9 ppm region. The experimental assignments from hairpin helices of the rT arm, the anticodon arm and the DHU arm, together with the ring current-predicted assignments of the acceptor stem base pairs, are indicated by their base-pair numbers on the abscissa. With one or two exceptions the individual assignments agree with the spectrum predicted by calculating the shifts for the tRNA$_1^{Val}$ sequence based on the tRNA^{Phe} crystal coordinates (Robillard $et\ al.$, 1976, 1977a, b).

ADDITIONAL RESONANCES FROM TERTIARY AND NONSTANDARD SECONDARY BASE PAIRS

The foregoing analysis of cloverleaf Watson-Crick base pairs leads to the conclusion that resonances between −11 ppm and −11.5 ppm and between −14 ppm and −15 ppm, as well as the protons at −13.3 ppm, −12.9 ppm, −12.2 ppm and −11.95 ppm, are not derived from the 20 Watson-Crick secondary pairs. These peaks are indicated by asterisks in figure 6.7. It cannot immediately be assumed that all these additional resonances are necessarily derived from tertiary base pairs since $E.\ coli$ tRNA$_1^{Val}$ contains a nonstandard GU pair in the rT stem at position 50 which may generate low field proton resonances. In support of this possibility anomalous peaks were noted in the spectrum of the GU-containing isolated rT hairpin helix (Reid $et\ al.$, 1979). To investigate GU pairs use was made of a nuclear Overhauser observation discovered by Johnston and Redfield (1979). The nuclear Overhauser effect (NOE) involves the transfer of saturation via dipolar coupling from a proton to adjacent protons in the immediate vicinity. The NOE is extremely distance dependent (r^{-6}) so that even ring NH protons of adjacent base pairs are too far apart for this process to be detectable. However, in the low field spectrum of yeast tRNA^Phe there are two ring NH protons which do exhibit reciprocal NOEs, indicating that they are extremely close together. Johnston and Redfield have shown that they are the ring NH protons of G and U *in the same*

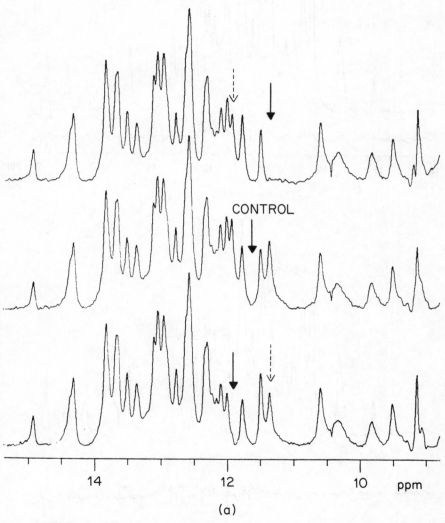

(a)

Figure 6.8 (a) Identification of the two ring NH protons of the GU pair in *E. coli* tRNA_1^Val by means of the NOE. The middle spectrum is the control preirradiated in the valley at −11.6 ppm. The upper spectrum shows the effect of 100 ms preirradiation (to saturation) of the peak at −11.35 ppm; note the cross-relaxation of the peak at −11.95 ppm (broken arrow). The lower spectrum shows the reciprocal saturation transfer when the peak at −11.95 ppm is pre-irradiated.

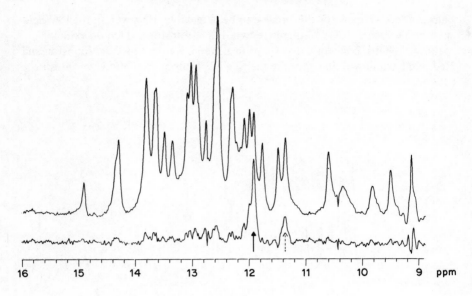

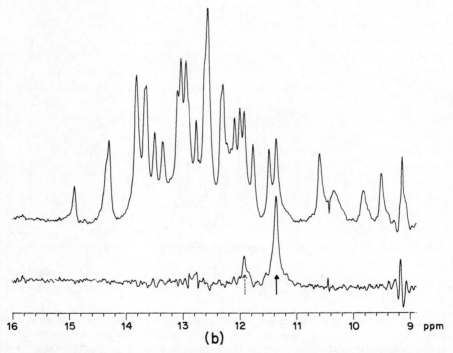

(b)

Figure 6.8 (b) The data of Figure 6.8a replotted as the control-minus-irradiated difference spectra. The observed NOE identifies peaks Q and T in Figure 6.7 as being the protons of GU50 *base pair*, that is the GU pair at position 4 in yeast tRNA^{Phe}. We have used this method to search for the GU pair at position 50 in *E. coli* tRNA$_1^{Val}$ and some results are shown in figure 6.8a, b. The effect on the 11.95 ppm peak of irradiating at −11.35 ppm and the effect on the 11.35 ppm peak of irradiating at −11.95 ppm, as well as the control irradiated at −11.6 ppm are shown in figure 6.8a. The reciprocal effects are better shown in the control minus irradiated difference spectra in figure 6.8b; transfer of saturation to the extent of about 30 per cent is evident between these two resonances under these conditions (100 ms preirradia-tion with a 1 ms delay followed by the detection pulse). This same result has been obtained for *E. coli* tRNA$_1^{Val}$ by Johnston and Redfield (personal communication) and since no other pair of low field protons exhibits reciprocal NOEs the 'extra' resonances at −11.95 ppm and −11.35 ppm in figure 6.7 must be the two ring NH protons of the GU 'wobble' base pair (the former is probably U64 and the latter G50 from the chemical shifts of the respective monomers). The reader may have realised at this point that this experiment could not be carried out using CW spectroscopy or correlation spectroscopy methods; the data in these, and subsequent preirradiation experiments, were obtained using long pulse Redfield 214 FTNMR methods (described on p. 133).

TERTIARY RESONANCES

The assignment of the Watson-Crick secondary base pairs and the two resonances from the wobble GU pair leaves six remaining resonances between −11 ppm and −15 ppm which must be assigned to tertiary base pairs in the three-dimensional folding of the molecule in solution. We have no method of determining *a priori* the origin of these tertiary interactions and progress in assigning these remaining tertiary resonances has relied heavily on the x-ray crystal structure of tRNA and the assumption that the solution structure is similar to the crystal structure. In the refined crystal structure there are seven or eight tertiary base pairs which involve ring NH hydrogen bonds; these are U8-A14 (reversed Hoogsteen), T54-A58 (reversed Hoogsteen), m^{7}G46-G22 (base triple), G19-C56 (Watson-Crick), ψ55-phosphate58 (base-backbone), m$_2^2$G26-A44 (purine-purine with propeller twist) and possibly G18-ψ55 (this hydrogen bond may be from the exocyclic amino group of G18 rather than the ring NH).

A variety of approaches have been taken in assigning the remaining six 'extra' resonances in figure 6.7.

The −14.9 ppm resonance is s^{4}U8-A14

E. coli tRNAs containing s^{4}U at position 8 contain a nonsecondary resonance at approximately −14.9 ppm whereas yeast tRNAs (U at position 8) do not (Reid *et al.*, 1975). Conversion of s^{4}U8 to U8 in *E. coli* tRNA$_1^{Val}$ shifts the −14.9 ppm resonance upfield to −14.3 ppm (Reid *et al.*, 1975; Wong *et al.*, 1975). Photo-oxidation of s^{4}U8 to U8 in *E. coli* tRNA^{fMet} causes the −14.9 ppm resonance to shift to -14.3 ppm (Daniel and Cohn, 1975). *E. coli* tRNA$_1^{Val}$ samples containing 50 per cent s^{4}U8 and 50 per cent U8 exhibit a half proton at -14.9 ppm and a half proton at -14.3 ppm (Hurd and Reid, 1979*b*). Together these observations assign the s^{4}U8 ring NH (presumably hydrogen bonded to A14 as in the crystal structure) at -14.9 ppm. The U8-A14 proton in yeast tRNA^{Phe} resonates at -14.3 ppm. The extremely low field chemical shift of this resonance in *E. coli* tRNAs is a result of additional deshielding of N3H by the adjacent sulphur in s^{4}U (Hurd and Reid, 1979*a*).

The -14.3 ppm resonance is probably T54-A58

Hydrogen bonds from GC pairs resonate upfield of -13.4 ppm and there is only one other crystallographic AU-type tertiary pair to account for the extra resonance in figure 6.7 at -14.3 ppm, namely T54-A58. This tertiary pair connects two residues in the same loop and can potentially be formed in isolated fragments. At lower temperatures the tRNA$_1^{Val}$ fragment 47-76 containing the T helix and T loop reveals an extra resonance at -14.3 ppm which is rapidly lost upon raising the temperature (Hurd and Reid, 1979). The resonance at -14.3 ppm in intact tRNA$_1^{Val}$ is perhaps the most labile proton in the whole molecule when the temperature is raised (Reid *et al.*, 1979). Upon removal of magnesium by EDTA the -14.3 ppm and -14.9 ppm resonances both show the rapid helix-coil breathing rates characteristic of tertiary base pairs (Johnston and Redfield, personal communication).

The -13.3 ppm resonance is m^{7}G46-G22

Chemical removal of m^{7}G46 under mild conditions eliminates the -13.3 ppm resonance from the spectrum of *E. coli* tRNA$_1^{Val}$ and other tRNAs containing m^{7}G46 in the same sequence environment (Hurd and Reid, 1979*b*). The extra resonance at -13.3 ppm is absent in the spectra of *E. coli* tRNA$_1^{Gly}$ and tRNA^{Gln} (which do not contain m^{7}G).

The -12.9 ppm resonance is G19-C56

The G19-C56 pair is the only tertiary interaction that is a normal Watson-Crick base pair. This base pair, which connects the D loop and T loop, is located at the extremity of the crystal structure and is only stacked on one side. The estimated ring current shift from proximal bases is approximately 0.5 ppm and if we assign an inherent intrinsic chemical shift of -13.45 ppm found for other Watson-Crick GC pairs this leads to an estimated chemical shift of approximately -12.95 ppm for G19-C56. Every tRNA we have examined contains G and C residues in these equivalent positions and all exhibit a tertiary resonance in the -12.8 to -13.0 ppm region. We know of no chemical methods for selectively modifying these residues and the assignment of this tertiary base pair must remain purely deductive.

The -12.3 ppm resonance is G15–C48 (in *E. coli* tRNA$_1^{Val}$)

This resonance was assigned from a study of the paramagnetic relaxation effects of cobalt bound to G15 and manganese bound to P8 (Hurd *et al.*, 1979). Both of these paramagnetic cation sites are close to s^{4}U8-A14 and G15-C48 and both cations selectively broaden the -14.9 ppm and -12.3 ppm resonance. The assignment is corroborated by a study of *E. coli* tRNA$_1^{Gly}$ which, instead of G and C, contains A and C in these equivalent positions; the -12.3 ppm resonance disappears in this spectrum since neither A nor C contain a ring NH proton.

The -11.5 ppm resonance is probably ψ55-P58

This is a weak deductive assignment which has not been demonstrated experimentally. The reasons for making this guess are the constancy of this pseudouridine residue in all tRNAs studied and the invariant extra resonance at -11.5 ppm. Although Pu26-Pu44 is a possible candidate in yeast tRNA^{Phe} and *E. coli* tRNA$_1^{Val}$ the -11.5 ppm tertiary resonance is always observed even in tRNA species with base substitutions in the 26 and 44 positions. Furthermore the observed position of this resonance at the high field end of the low field spectrum is consistent with the loss of the downfield shift upon hydrogen bonding to a phosphate acceptor instead of a base acceptor. Despite these rationalisations this assigment is still only an inspired guess at best.

From such an analysis the conclusion is that the Pu26-Pu44 ring NH proton (A26-G44 in *E. coli* tRNA$_1^{Val}$ and m$_2^2$G26-A44 in yeast tRNA^{Phe}) and the G18-ψ55 ring NH hydrogen bond (if it exists) are not detectable in the -11 ppm to -15 ppm spectrum. It is possible that these protons resonate further upfield in the -9 ppm to -11 ppm region due to very weak hydrogen bonding; an alternative possibility is that these protons are not detectable due to excessively short helix lifetimes. A tacit assumption throughout the foregoing analysis is that the spectrum of *E. coli* tRNA$_1^{Val}$ can be interpreted on the basis of the structure of yeast tRNA^{Phe}. This assumption is probably justifiable because the tertiary interaction positions 8, 14, 15, 19, 46, 48, 54, 56, 58, and so on, in the latter tRNA are occupied by the same residues in the primary sequence of the former molecule. Such analyses of allowed positional substitutions have led to the proposal that such closely related tRNA species fold into similar tertiary structures (Klug *et al.*, 1974; Kim *et al.*, 1974).

HELIX DYNAMICS IN tRNA

The hydrogen-bonded ring NH protons are buried in the interior of RNA helices where they are inaccessible to the surrounding water. However, they do exchange with solvent protons on a reasonable time scale. The most reasonable model for helix–water exchange is via the open coil state as shown in figure 6.9. The mechanism of this exchange process has been analysed in detail by Teitelbaum and Englander (1975*a, b*). At 0°C k_{open} is a relatively slow process (seconds) whereas

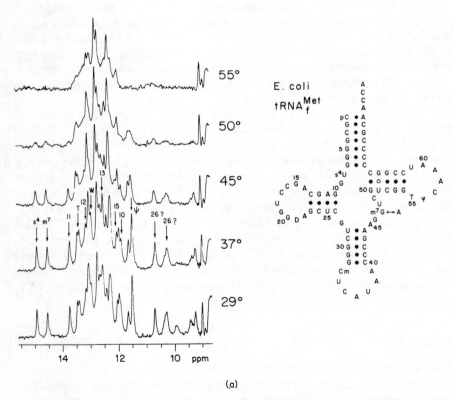

Figure 6.9 Diagrammatic representation of the mechanism of exchange of a helical ring NH proton with water via the coil state. When the coil-water rate is made very rapid, compared to k_{open} and k_{close}, exchange is open limited. When the helix opening rate is fast compared to the purely magnetic spin-lattice relaxation rate the recovery of a resonance which has been pre-irradiated to saturation will be exchange dominated.

(a)

Figure 6.10 (a) NMR analysis of the thermal unfolding sequence of *E. coli* tRNA$_f^{Met}$ in 1 mM sodium phosphate, 10 mM sodium cacodylate, 5 mM EDTA, 144 mM NaClO$_4$ pH 7.0 (total sodium is 0.174 M). The buffer is identical to that described by Crothers *et al.* (1974). The optical T_m for tertiary structure under these conditions is 45°; note the resonances below −13.5 ppm which have lost 50 per cent intensity by 45° without appreciable line broadening.

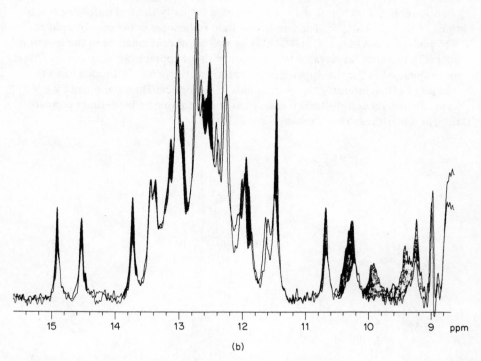

Figure 6.10 (b) Superimposition of the 29° and 45° spectra of figure 6.10a. The intensity losses have been shaded for ease of comparison.

k_1 [cat] is much more rapid (and depends on the pK and concentration of the buffer used), hence the helix–water exchange rate is limited by k_{open}. At or above physiological temperature k_{open} is presumably much faster than at $0\,^{\circ}$C and is no longer experimentally amenable by the tritium gel filtration methods developed by Englander and colleagues. However, NMR methods are ideally suited to probing such processes in the millisecond time range and there are two alternative approaches to studying helix-coil dynamics. The first approach makes use of the previously mentioned requirement for an approximately 5 ms (or longer) helix lifetime in order to observe a narrow low field ring NH signal. The various helices in tRNA exist at physiological temperature because k_{open} is still much slower than k_{close} and the observation of narrow lines indicates that k_{open} (the reciprocal helix lifetime) must be less than 200 sec^{-1}. At some elevated temperature, i.e. the thermodynamic Tm, the accelerated k_{open} will equal k_{close} which varies depending on the helix in question. This can happen either before or after the 200 s^{-1} (5 ms lifetime) 'time window' has been attained. If, for example, $k_{close} = 40$ s^{-1} then at the Tm $k_{open} = k_{close}$ and the helix lifetime is 25 ms which is long enough to generate a narrow low field resonance. However, at any instant, 50 per cent of the population is in the coil state (from which the proton is lost by rapid water exchange) so that the helix resonance, while still narrow, will only have an intensity of 0.5. An example of this phenomenon is shown in figure 6.10a, b during the early thermal unfolding transitions of *E. coli* tRNA$_f^{Met}$. The two lowest field resonances between −14 ppm and −15 ppm (s^{4}U8-A14 and m^{7}G46-G22), as well as other resonances in the spectrum, gradually disappear at elevated temperature without appreciable broadening. This is more apparent in the two superimposed spectra in figure 6.10b. In such cases the mid-point of the intensity loss corresponds to the optical *Tm* determined by UV hyperchromicity and the lines remain narrow due to long helix lifetimes even at the *Tm* under these 'slow exchange' conditions.

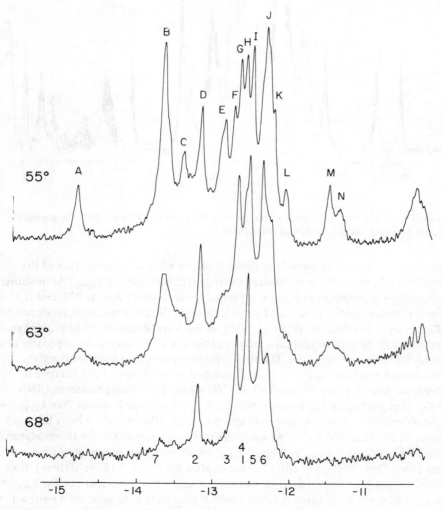

Figure 6.11   The NMR spectrum of *E. coli* tRNA^{Phe} at intermediate temperatures during the thermal unfolding in the presence of magnesium. Note the linebroadening of the s^{4}U8 resonance (peak A).

If, on the other hand, k_{close} is much faster than 200 s^{-1}, say for example 1000 s^{-1}, then a quite different effect is observed for the low field resonances as the temperature is gradually increased. In this case k_{open} will gradually increase, but well before it reaches k_{close} it will pass through the 200 s^{-1} time window at which point it will exchange-broaden and disappear from the spectrum. In this case the NMR 'melting' temperature will be several degrees lower than the thermodynamic Tm determined optically. An example of this process is shown in figure 6.11 for the case of $E.$ $coli$ tRNA^{Phe} in the presence of magnesium. At 55 °C the isolated resonance from s^{4}U8-A14 has already begun to broaden compared to its linewidth at 37 °C. By 63 °C this resonance, as well as other peaks L, M, N, and so on, has undergone serious exchange broadening and has disappeared by 68 °C. The relevant exchange parameters can be determined with moderate accuracy from linewidth analysis and the underlying theory has been presented by Crothers et $al.$ (1974).

TIME-RESOLVED FTNMR: INDIVIDUAL BASE-PAIR BREATHING RATES

Although the above linewidth and intensity methods are useful in the analysis of helix kinetic parameters in fast-exchange or slow-exchange conditions near the Tm, they do not reveal kinetic information about helix-coil dynamics at intermediate temperatures where linewidths are at the narrow limit and insensitive to exchange rate. It is obvious that these kinetic parameters in the millisecond time range cannot be studied by correlation spectroscopy or by CW spectroscopy where a single sweep of a time-averaged spectrum may take a second or several seconds respectively. Investigation of processes in the millisecond range requires the time-resolution of pulsed FTNMR yet it is equally obvious that FTNMR of tRNA-H$_2$O systems is beset by enormous computer dynamic range problems involving signal-to-noise ratios in spectra containing 1 mM solute protons in the presence of 110 M solvent proton peaks. These problems have been discussed by Redfield (1978) and elegantly solved by the introduction of the Redfield 214 detection pulse sequence. The advantages of this technique are illustrated conceptually in figure 6.12, in which the 'spectrum' refers to the low field tRNA ring NH protons of interest. In normal FTNMR it is usual to choose a detection pulse of approximately 10 μs generating a broad frequency distribution so that the amplitude of the frequencies across the region of interest is virtually constant. The Fourier transform of a square pulse drops to zero at the reciprocal of the pulse length; hence the effect of a 500 μs long pulse whose frequency is offset by 2000 Hz from water is a drastic reduction in the excitation of water protons. The amplitude at the solvent frequency can be further reduced by subdividing the long pulse into consecutive tenths, the third and eighth of which are phase inverted; this is the Redfield 214 pulse sequence which reduces the effective water peak height to below 1 M thus placing 1 mM resonances within the dynamic range of most digitisers. The two most important uses of the Redfield 214 pulse technique are that solvent protons are unperturbed and that the detection pulse takes less than a millisecond so that processes in the 1–1000 ms range can be monitored using preirradiation methods. Johnston and

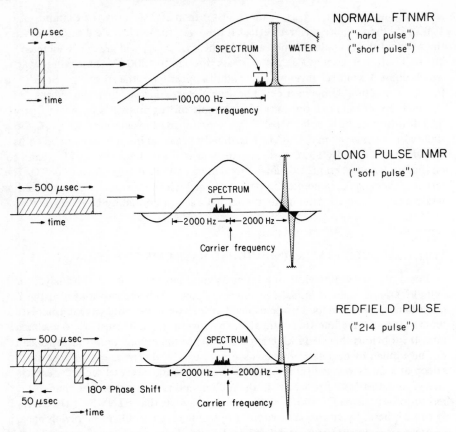

Figure 6.12 Conceptual representation (not to scale) of standard, long and Redfield 214 pulse FTNMR methods in H_2O solution.

Redfield (1977, 1978) have used this method to study base-pair exchange rates in yeast tRNA[Phe]. Referring back to the model in figure 6.9, if the ring NH resonances of these three base pairs are resolved (and assigned) in the low field spectrum we now have the opportunity to selectively irradiate to saturation only the middle base pair using, for instance, a 100 ms preirradiation (10–20 Hz monochromatic rf) tuned to this specific peak. After switching off the saturating preirradiation various time intervals can be allowed to elapse before applying the 214 detection pulse and thus determining the rate at which this proton recovers from saturation (the experiment is analogous to the second half of a classical 180°-tau-90° T_1 spin-lattice relaxation experiment since saturation can be considered as the absence of z magnetisation). In contrast to the purely magnetic spin-lattice recovery (T_1^{-1}) of non-exchangeable CH protons, the ring NH protons have an additional method of saturation recovery, namely chemical exchange with a fresh water proton (as opposed to a marine proton?) which is not saturated since it was never preirradiated.

When k_1 [cat] $\gg k_{open}$, k_{close}, the recovery rate equation can be simplified to the form:

$$\frac{dH}{dt} = -\frac{H}{\tau_{hw}} + \frac{H^\circ}{\tau_{hw}} + \frac{H^\circ - H}{T_{1h}}$$

where H° is the full proton intensity (infinite recovery), $1/\tau_{hw}$ is the helix-water exchange rate, and $1/T_{1h}$ is the magnetic spin-lattice relaxation rate. Hence a plot of the logarithm of relative intensity as a function of the recovery time delay before the detection pulse has the slope $(1/\tau_{hw} + 1/T_{1h})$ for the observed recovery rate. Under buffer-catalysed conditions (around 10 mM buffer) at neutral pH k_1 [cat] is very rapid (10^4–10^5 s^{-1} depending on the buffer pK) compared to k_{open} or k_{close} and helix–water exchange is rate-limited by k_{open}. Furthermore, when exchange is rapid compared to spin-lattice relaxation the experimentally observed recovery rate approximates k_{open} directly, that is, the measured slope reveals the base pair breathing rate when saturation recovery is exchange dominated. Although it has not been measured directly, our best estimates for the spin-lattice relaxation rate $1/T_{1h}$ are in the range of 5–7 s^{-1}; hence observed recovery rates in the range of 50–100 s^{-1} are dominated to the extent of 90–95 per cent by k_{open} and in these cases the slope of the recovery plot very closely approximates the base-pair opening rate since there is at least 5 per cent experimental error in determining the peak intensity at each delay time in spectra with signal-to-noise ratios of approximately 20.

An example of this type of analysis of helix-coil dynamics for single base pair in a defined helix is shown in figure 6.13a, b. The helix chosen was the GC-CG-CG-CG-GC-GC acceptor helix of *E. coli* tRNA[Phe] which is the only remaining helix in this molecule detectable by NMR at elevated temperatures. Although it is necessary to go to 68° to exchange out the other 21 base pairs in the presence of magnesium, this can be accomplished by 52° in phosphate-NaCl buffers in the absence of magnesium. Under these latter conditions only resonances from base pairs 1, 2, 3, 4, 5, 6 in the temperature range from 52° to 62° are observed, thus permitting analysis of the temperature dependence of base-pair opening and determination of the activation energy. Figure 6.13a shows the effect of preirradiating CG2 at −13.18 ppm and monitoring its saturation recovery at various time intervals at two temperatures. Figure 6.13b shows the data in plotted form from which the slopes at three temperatures can be determined. These rates are 14 s^{-1}, 25 s^{-1} and 45 s^{-1} at 52°, 58° and 62 °C, respectively. These values cannot be directly equated with k_{open} since the experimentally observed recovery rate is not heavily dominated by exchange and a significant component of this rate (approximately 6 s^{-1}) is probably magnetic spin-lattice recovery. Similar analyses have been carried out on the other five base pairs in this helix (Hurd and Reid, unpublished data) and the results can only be interpreted in terms of a model in which the helix frays and breathes rapidly at both ends. This approach to studying the dynamics of helix-coil transitions for individual base pairs is an extremely promising tool for studying the solution conformation of intact tRNA (Johnston and Redfield, 1977, 1978) and other RNA molecules with reasonably resolved low field spectra.

This chapter has discussed experimental approaches to probing the structure and dynamics of tRNA in solution utilising the extreme low field ring NH hydrogen-bond NMR spectrum exclusively. Although the spectrum of this polymer between −4 ppm and −9 ppm is too complex and unresolved for assignment and detailed interpretation, the extreme high field NMR spectrum of tRNA between 0 ppm and −4 ppm contains several resolved and usefully diagnostic proton resonances for structural studies. These resonances are the aliphatic protons of modified nucleotides such as m^6A, Y, t^6A, T, DHU, m^5C, and so on. A disadvantage in this spectral region is the inability to study exchange in these aliphatic carbon protons which restricts these studies to changes in chemical shift, linewidth and T_1. However, a compensating advantage in the high field spectrum is the fact that the nucleotides generating these resonances are usually located in the loop regions rather than the helical regions of tRNA. The chemical shift information derived from such unpaired residues in tRNA loops is often useful in complementing helical data from the low

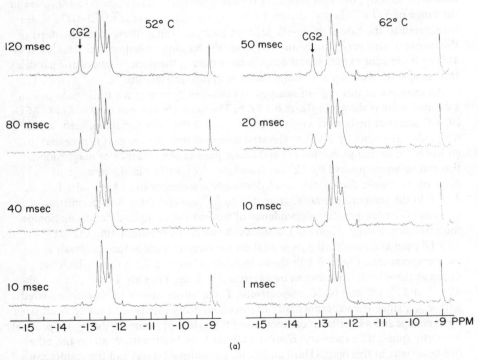

(a)

Figure 6.13   (a) Experimental determination of the helix-coil breathing rate of CG2 in the acceptor helix of *E. coli* tRNA^{Phe} by saturation recovery analysis. Selective preirradiation was carried out for 100 ms at −13.12 ppm. At various time delays after switching off the pre-irradiation, the spectrum was acquired with a Redfield 214 detection pulse. The effect of temperature and the lack of effect of pH or additional buffers confirmed that exchange was open limited and that saturation recovery was exchange dominated.

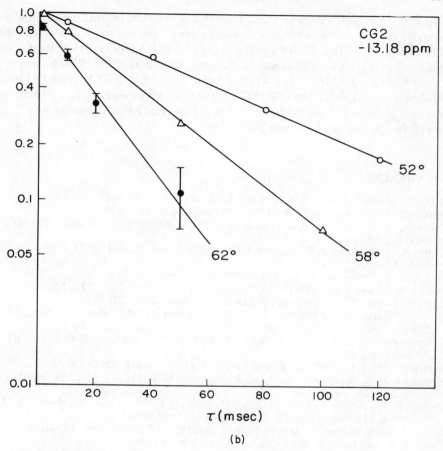

Figure 6.13   (b) Semilogarithmic plot of the saturation recovery rates of CG2 at three temperatures.

field spectrum. Several studies of tRNA high field spectra have been reported in the recent literature which the reader is encouraged to read (Kan *et al.*, 1974; Kan *et al.*, 1977; Robillard *et al.*, 1977*a*; Kastrup and Schmidt, 1975; Kastrup and Schmidt, 1978).

ACKNOWLEDGEMENTS

The 360 MHz spectra in this article were obtained on a modified Bruker HXS 360 spectrometer at Stanford University Magnetic Resonance Laboratory (supported by NSF Grant GR 23633 and NIH Grant RR00711). The author thanks the SMRL committee for machine time and W. W. Conover for instrument modifications.

Thanks are also due to students Ralph Hurd, Ed. Azhderian, Do-lan Hoo and Kathy Jones for stimulating discussion and the use of some of their research data and to Susan Ribeiro and Lillian McCollum for isolating the pure tRNA species used in these studies. During the course of this work the research support of the NIH-National Cancer Institute (CA11697), the National Science Foundation (PCM 7902904) and the American Cancer Society (NP-191B) is gratefully acknowledge. The 600 MHz spectrum in figure 6.2 was kindly run for us at Carnegie-Mellon University by Joseph Dadok and Aksel Bothner-By.

REFERENCES

Arter, D. B. and Schmidt, P. G. (1976). *Nucl. Acids Res.*, **3**, 1437.

Bolton, P. H. and Kearns, D. R. (1976). *Nature, Lond.*, **262**, 423.

Crothers, D. M. and Cole, P. E. (1978). In *Transfer RNA* (ed.), S. Altman, The MIT Press, Cambridge, MA, p. 196.

Crothers, D. M., Cole, P. E., Hilbers, C. W. and Shulman, R. G. (1974). *J. Mol. Biol.*, **87**, 63.

Dadok, J. and Sprecher, R. F. (1974). *J. Magn. Reson.*, **13**, 243.

Daniel, W. E. and Cohn, M. (1975). *Proc. natn. Acad. Sci. U.S.A.*, **72**, 2582.

Giessner-Prettre, C. and Pullman, B. (1970). *J. Theor. Biol.*, **27**, 87.

Giessner-Prettre, C. and Pullman, B. (1976). *Biochem. biophys. Res. Commun.*, **70**, 578.

Giessner-Prettre, C., Pullman, B., Borer, P. N., Kan, L. S. and Ts'o, P. O. P. (1976). *Biopolymers*, **15**, 2277.

Giessner-Prettre, C., Pullman, B. and Caillet, J. (1977). *Nucl. Acids Res.*, **4**, 99.

Hoagland, M. B., Zamecnik, P. C. and Stephenson, M. L. (1957). *Biochim. biophys. Acta*, **24**, 215.

Holley, R. W., Apgar, J., Everett, G. A., Madison, J. T., Marquisee, M., Merrill, S. H., Penswick, J. R. and Zamir, A. (1965). *Science*, **147**, 1462.

Hurd, R. E., Azhderian, E. M. and Reid, B. R. (1979). *Biochemistry*, **18**, 4012.

Hurd, R. E. and Reid, B. R. (1979a). *Biochemistry*, **18**, 4005.

Hurd, R. E. and Reid, B. R. (1979b). *Biochemistry*, **18**, 4017.

Jack, A., Ladner, J. E. and Klug, A. (1976). *J. molec. Biol.*, **108**, 619.

Johnston, P. D. and Redfield, A. G. (1977). *Nucl. Acids Res.*, **4**, 3599.

Johnston, P. D. and Redfield, A. G. (1978). *Nucl. Acids Res.*, **5**, 3913.

Johnston, P. D. and Redfield, A. G. (1979). To be published.

Jones, C. R. and Kearns, D. R. (1975). *Biochemistry*, **14**, 2660.

Kan, L. S., Ts'o, P. O. P., von der Haar, F., Sprinzl, M. and Cramer, F. (1974). *Biochem. biophys. Res. Commun.*, **59**, 22.

Kan, L. S., Ts'o, P. O. P., Sprinzl, M., von der Haar, F. and Cramer, F. (1977). *Biochemistry*, **16**, 3143.

Kastrup, R. V. and Schmidt, P. G. (1975). *Biochemistry*, **14**, 3612.

Kastrup, R. V. and Schmidt, P. G. (1978). *Nucl. Acids Res.*, **5**, 257.

Katz, L. and Penman, S. (1966). *J. molec. Biol.*, **15**, 220.

Kearns, D. R. (1976). *Prog. Nucl. Acid. Res. Mol. Biol.*, **18**, 91.

Kearns, D. R., Patel, D. J., Shulman, R. G. and Yamane, T. (1971). *J. molec. Biol.*, **61**, 265.

Kearns, D. R. and Shulman, R. G. (1974). *Acc. Chem. Res.*, **7**, 33.

Kim, S. H., Suddath, F. L., Quigley, G. J., McPherson, A., Sussman, J. L., Wang, A., Seeman, N. C. and Rich, A. (1974). *Science*, **185**, 435.

Kim, S. H., Sussman, J. L., Suddath, F. L., Quigley, G. J., McPherson, A., Wang, A., Seeman, N. C. and Rich, A. (1974). *Proc. natn. Acad. Sci. U.S.A.*, **71**, 4970.
Klug, A., Ladner, J. and Robertus, J. D. (1974). *J. molec. Biol.*, **89**, 511.
MacDonald, C. C., Phillips, W. D. and Penman, S. (1964). *Science*, **144**, 1234.
Quigley, G. J., Wang, A. H. J., Seeman, N. C., Suddath, F. L., Rich, A., Sussman, J. L. and Kim, S. H. (1975). *Proc. natn. Acad. Sci. U.S.A.*, **72**, 4866.
Redfield, A. G. (1978). *Methods Enzymology*, **49**, 253.
Reid, B. R., McCollum, L., Ribeiro, N. S., Abbate, J. and Hurd, R. E. (1979). *Biochemistry*, **18**, 3996.
Reid, B. R., Ribeiro, N. S., Gould, G., Robillard, G., Hilbers, C. W. and Shulman, R. G. (1975). *Proc. natn. Acad. Sci. U.S.A.*, **72**, 2049.
Reid, B. R., Ribeiro, N. S., McCollum, L., Abbate, J. and Hurd, R. E. (1977). *Biochemistry*, **16**, 2086.
Robertus, J. D., Ladner, J. E., Finch, J. T., Rhodes, D., Brown, R. D., Clark, B. F. C. and Klug, A. (1974). *Nature, Lond.*, **250**, 546.
Robillard, G. T., Tarr, C. E., Vosman, F. and Berendsen, H. J. C. (1976). *Nature, Lond.*, **262**, 363.
Robillard, G. T., Tarr, C. E., Vosman, F. and Reid, B. R. (1977a). *Biochemistry*, **16**, 5261.
Robillard, G. T., Tarr, C. E., Vosman, F., Sussman, J. L. and Kim, S. H. (1977b). *Biophys. Chem.*, **6**, 291.
Smith, I. C. P., Yamane, T. and Shulman, R. G. (1968). *Science*, **159**, 1360.
Smith, I. C. P., Yamane, T. and Shulman, R. G. (1969). *Can. J. Biochem.*, **47**, 480.
Sprinzl, M., Gruter, F. and Gauss, D. H. (1978). *Nucl. Acids Res.*, **5**, r15.
Stout, C. D., Mizuno, H., Rao, S. T., Swaminathan, P., Rubin, J., Brennan, T. and Sundaralingam, M. (1978). *Acta crystallogr.*, **B34**, 1529.
Sussman, J. L., Holbrook, S. R., Wade-Warrant, R., Church, G. M. and Kim, S. H. (1978). *J. molec. Biol.*, **123**, 607.
Teitelbaum, H. and Englander, S. W. (1975a). *J. molec. Biol.*, **92**, 55.
Teitelbaum, H. and Englander, S. W. (1975b). *J. molec. Biol.*, **92**, 79.
Wong, K. L., Bolton, P. H. and Kearns, D. R. (1975). *Biochim. biphys. Acta.*, **383**, 446.

7

Eukaryotic chromatin

D. M. J. Lilley

INTRODUCTION

The structure of DNA in its functional state in cells is modified by protein binding and the introduction of superhelix density. Whilst this is probably true for most natural DNA molecules, it is especially so for eukaryotic DNA, where an at least equal weight of protein serves to contract about one metre of nucleic acid by approximately five orders of magnitude. Typically, a human somatic cell contains about one thousand times the amount of DNA present in prokaryotic cells, where it is condensed into a nucleus of diameter 5 μm whilst allowing access to the enzymes involved in all aspects of nucleic acid metabolism. In eukaryotes discrete DNA molecules of between 10^4 to 10^6 kilobase pairs (Kb) in length are organised by protein into morphologically and genetically distinct units called chromosomes. The nucleoprotein material of which they are composed is generally called chromatin.

The composition of eukaryotic chromatin is variable, but the major components, DNA and histone, are almost universally present as a one to one complex by weight. The histones are a class of small basic proteins which may be divided into the core histones, H2A, H2B, H3 and H4, and the lysine-rich histone H1 and similar species-specific molecules. The crucial role of these proteins in chromatin packaging is emphasised by their evolutionary sequence conservation. The reason for the sub-division of the histones will later be apparent, but there are excellent genetic reasons for considering the five histones as a single class of molecules. Histones may be modified *in vivo* by methylation, acetylation and phosphorylation at specific positions in their primary sequences, and genetic variants of the histones appear to be expressed under developmental control (Newrock *et al.*, 1978). The remaining proteins of eukaryotic chromatin are generally embraced in the term non-histone proteins (NHP). This is a complex and heterogeneous class of protein which contains enzymes, structural proteins and putative regulatory species. Some members of this group, such as the high mobility group (HMG) protein (Johns *et al.*, 1975), are present in relatively large quantities and may have an important structural role.

Significant progress has been made in the last five years in understanding how the histones organise DNA into chromatin structure at a molecular level. More recently some first steps have been taken in elucidating the more dynamic and functional aspects of the structure. It is the purpose of this chapter to review these successes and to make some prediction of how this field will progress. It will be appreciated that this is a rapidly moving area of molecular biology and by the time this review is published further important advances may well have been made. It should, however, serve as a detailed introduction to this fascinating and important area of molecular genetics.

CHROMATIN AS A SUBUNIT STRUCTURE

The first appreciation that DNA might be organised into a chromatin fibre in a discontinuous manner came from the work of Olins and Olins (1974), who examined chromatin from osmotically ruptured avian erythrocyte nuclei by electron microscopy. Instead of visualising a fibre of constant diameter they observed bead-like structures around 70 Å diameter spaced along the DNA. Such a regularly repeating subunit type of structure, likened to 'beads on a string', could explain observations (Hewish and Burgoyne, 1973; Noll, 1974a; Shaw et al., 1976) that nucleases cleave chromatin at intervals of about 200 base pairs (bp) along the DNA. Kornberg and Thomas (1974) demonstrated that histones can be isolated at discrete stoichiometric units, and two proposals were made (Kornberg, 1974; Van Holde, Sahasrabuddhe and Shaw, 1974) that an octamer containing two molecules each of histones H2A, H2B, H3 and H4 organises approximately 200 bp of DNA into the basic chromatin subunit, now generally termed a nucleosome.

Careful analysis of micrococcal nuclease digestion kinetics (Lohr et al., 1977; Noll and Kornberg, 1977) indicates that although the overall DNA repeat length in chromatin is around 200 bp, further nuclease action progressively 'trims' the DNA ends of the monomeric subunit, with kinetic pauses at around 160 and 145 bp. In this way a nuclease resistant 'core particle' containing 145 bp DNA and the histone octamer has been identified. No lysine rich histone (H1 or similar molecules) are associated with this particle, indicating that this histone is associated not primarily with the core itself but with the DNA connecting cores, sometimes called the spacer or linker DNA. As discussed later, it is probable that the thread of nucleosomes can be further compacted to generate a higher level of structure, for which H1 appears to have an important role.

All eukaryotes examined have been found to have chromatin organised as subunit structures (for reviews see Kornberg, 1977; Chambon, 1978; Felsenfeld, 1978; Van Holde and Weischet, 1978). Furthermore the core particles are closely similar in each case, around 145 bp of DNA in length. However, some variability is present in overall repeat lengths, that is in the length of the spacer DNA, some examples of which are listed in table 7.1. No systematic generalisations are apparent from these data but some general points can be made. Considerable variation in spacer length is found, from 9 bp in *Aspergillus* to 96 bp in sea urchin sperm. Whilst the fungi appear to have rather short repeat lengths, no other phylogenetic generali-

Table 7.1 Some representative repeat lengths measured for various eukaryotes. (More extensive tabulations may be found in Kornberg (1977) and Van Holde and Weischet (1978))

Cell	Repeat (bp)	Reference
Aspergillus	154	Morris (1976)
Yeast	162 ± 2	Thomas and Furber (1976), Lohr, Kovacic and Van Holde (1977)
Physarum	171	Compton, Bellard and Chambon (1976)
Rye	200	Cheah and Osborne (1977)
Sea urchin gastrula	217 ⎫	Spadafora *et al.* (1976)
Sea urchin sperm	242 ⎭	
Chicken erythrocyte	207	Compton *et al.* (1976)
Rat liver	196	Compton *et al.* (1976)
Rabbit cortical neurone	160 ⎫	
Rabbit glial cell	200 ⎭	Thomas and Thompson (1977)
HeLa cells	188	Compton *et al.* (1976)

sation seems possible. Indeed, different cell types from the same tissue have been found to have repeat lengths which vary by as much as 40 bp (Thomas and Thompson, 1977) and a natural heterogeneity in spacer length has been shown to be present in rat liver chromatin (Prunell and Kornberg, 1978). Transcriptionally inactive cell types such as terminally differentiated avian erythrocytes tend to exhibit longer spacer lengths, but this may not be significant. Active genes have been shown (Gottesfeld and Melton, 1978) to have the same repeat length as bulk chromatin within the same nuclei. A related question concerns the possible existence of nucleosome phasing, that is association of nucleosomes with specific location on the DNA sequence. This seems to have been excluded for viral chromosomes (Pfeiffer *et al.*, 1975), rat liver bulk chromatin (Prunell and Kornberg, 1978) and even for a specific gene (Garel and Axel, 1978). However, highly localised phasing is still possible. What then is the *raison d'etre* for spacer length variability? No obvious correlations with cell type or activity present themselves, yet it seems likely that spacer length will place constraints upon higher levels of chromatin fibre folding. A conclusive answer will probably come only with a more detailed understanding of higher order structure in chromatin.

THE CORE PARTICLE

Since the basic unit of chromatin structure is the core particle it is important to have a good understanding of its structure at the molecular level. Fortunately techniques are available for preparation and purification of these particles (Shaw *et al.*, 1976), resulting in solutions which are sufficiently monodisperse to allow investigation by physical methods. Figure 7.1 shows the purification of nucleosome core particles and oligomers from chicken erythrocyte nuclei digested with micrococcal nuclease. In addition, it is possible to isolate pure core protein, a discrete stoichiometric complex of histones H2A, H2B, H3 and H4 (Thomas and Kornberg, 1975; Weintraub, Palter and Van Lente, 1975) for comparative studies.

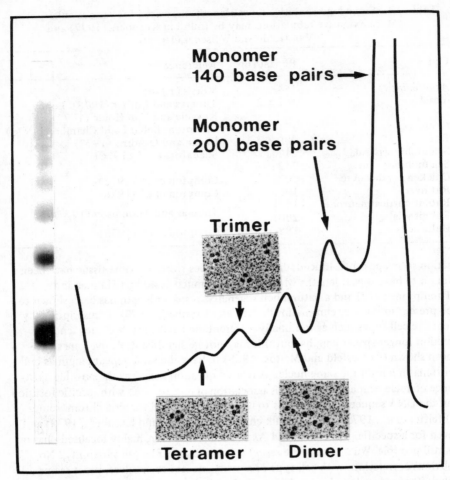

Figure 7.1   Chromatin is cleaved at sites separated by about 200 base pairs by nucleases. The DNA fragments, multiples of the 200 bp unit, may be separated by polyacrylamide gel electrophoresis, shown on the left. The corresponding DNA-histone species may be separated by sucrose gradient centrifugation, when peaks are produced indicating the separation of mono and oligonucleosomes, which may be thus purified. Electron micrographs of di, tri and tetranucleosomes are shown.

DNA and a histone core

In the original models proposed for a subunit chromatin structure (Kornberg, 1974; Van Holde *et al.*, 1974) it was suggested that the DNA would be wrapped around an octameric histone core. Evidence consistent with this proposal came from Noll (1974*b*) who showed that nucleosomal DNA was accessible to pancreatic DNase I at many locations within the particle.

Neutron scattering studies were able to provide more rigorous proof of this contention (Pardon et al., 1975; Hjelm et al., 1977). Neutrons are scattered by atomic nuclei and the scattering lengths for different nuclei vary over a wide range. In particular, hydrogen and deuterium scatter with opposite phase. The result of this is that the average scattering amplitudes for various biological materials are different, and are bracketed by the range of H_2O to D_2O. Thus by the nonperturbing substitution of D_2O for H_2O in varying proportion, the contrast of the different macromolecular components may be varied at will. By measuring the scattered neutron intensity from core particle solutions at low angles such that the Guinier law (Guinier and Fournet, 1955) is obeyed,

$$I_Q = I_o \exp(-\tfrac{1}{3} R_g^2 Q^2), Q = \frac{4\pi}{\lambda} \sin \theta \qquad (7.1)$$

where I_Q and I_o are the scattered and zero angle intensities, θ is the scatter angle and λ is the neutron wavelength, the radius of gyration, R_g, may be measured as a function of contrast. At larger scattering angles interference effects arising from the shape and internal scattering density fluctuations become significant, producing secondary maxima and minima in the scattering curve from which shape information may be derived.

Table 7.2 Radii of gyration of the nucleosome and its components and average scattering amplitudes for DNA, histone and solvent

Component	Average scattering amplitude $(10^{-14} \text{ cm Å}^{-3})$	Radius of gyration (Å) (Pardon et al., 1975)
H_2O	−0.55	
D_2O	+6.36	
Core particle:		
Protein dominant	3.11	30.6 ± 2.0
DNA dominant	4.44	50.5 ± 1.4
Infinite contrast		41.1 ± 0.4

It was shown for core particles (Pardon et al., 1975; Hjelm et al., 1977) that the measured radius of gyration depends markedly upon solvent contrast, indicating that DNA and histone occupy separate domains in the particle. The sign of the second moment of the scattering density distribution indicated that the DNA has a larger R_g than the histone, whilst the absence of quadratic dependence upon contrast required that (to within experimental error) the components share a common centre of mass. Thus the DNA is external to a histone core. By calculation of scattering densities for protein and DNA, their individual radii of gyration were obtained, as shown in table 7.2. The values of 30.6 Å and 50.5 Å, respectively, clearly demonstrates the presence of an external DNA domain in these particles. It should perhaps be emphasised that these R_g values are 'hard' numbers which must be accommodated in any subsequent model building.

The histone core

It is not possible to discuss core particle structure without mention of studies on the histone complexes which constitute the particle core. It is a fortunate conse-quence of the external localisation of the DNA on the nucleosome that it can be removed under mild conditions, such as high ionic strength, with relatively little dis-ruption of the underlying protein structure. In fact Stein, Bina-Stein and Simpson (1977) have shown that the DNA may be reversibly dissociated even from a chemically crosslinked histone core. Three main stoichiometric histone complexes have been prepared from chromatin: an H2A H2B dimer (Kelley, 1973), an (H3 H4)$_2$ tetramer (Van der Westhuyzen and Von Holt, 1971; Kornberg and Thomas, 1974; Roark, Geoghegan and Keller, 1974) and core protein, which con-tains all four core histones. The latter complex is prepared simply by dissociating chromatin in 2 M sodium chloride and removing the DNA by sedimentation (Thomas and Kornberg, 1975; Weintraub et al., 1975). It can be assumed that this species is closely related to the nucleosome core, but there has been some debate concerning whether it is an octamer, that is two each of the core histones (Thomas and Butler, 1977; Eickbush and Moudrianakis, 1978) or a tetramer (Weintraub et al., 1975; Campbell and Cotter, 1976). Whilst all the data supporting both conten-tions seem good, it should be pointed out that physical studies at high ionic strength are not easy, and are complicated by several artefacts. It is also possible that slight differences in preparative procedures result in different species, and an octamer–tetramer equilibrium has been proposed from one study (Chung, Hill and Doty, 1978).

Hydrodynamic studies of the histone complexes show clearly that they are discrete entities and high resolution ^{1}H nuclear magnetic resonance (NMR) investiga-tions (Moss et al., 1976a; Lilley, Pardon and Richards, 1977) show that they possess precise tertiary structure. A typical ^{1}H NMR spectrum of core protein is shown in figure 7.2. Ring current shifted methyl proton resonances, upfield from 1 ppm, and similarly shifted aromatic proton resonances upfield from 6.8 ppm and downfield from 7.2 ppm indicate close and defined spatial approximation of aromatic side chains with isoleucine, valine or leucine and tyrosine or phenylalanine residues such as would be found on the hydrophobic interior of a globular protein. In addition, the splitting of the histidine C_2H proton resonances at 7.9 ppm indicates variation in local environment within the core, thereby perturbing pK values for the imidazole side chains. In the case of H2A-H2B dimer, Moss et al. (1976b) have identified one histidine residue titrating with the very low pK of 4.9. D'Anna and Isenberg (1974) have made a detailed study of histone-histone interactions by spectroscopic studies and concluded that strong interactions exist between histones H3-H4, H4-H2B and H2B-H2A, with weaker interactions between other pairs. These interactions are sufficient to generate the observed quaternary structures, and Isenberg has further

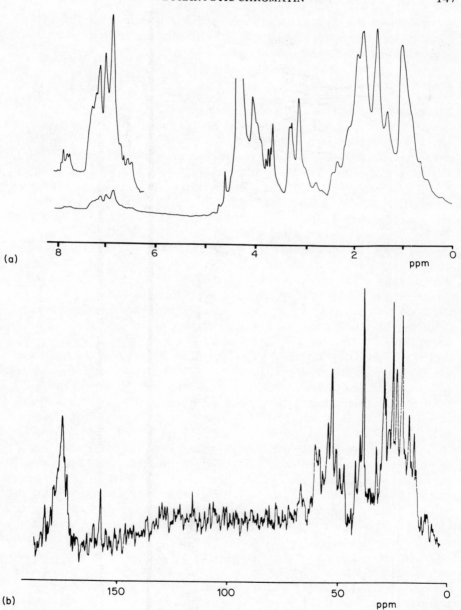

Figure 7.2   Nuclear magnetic resonance spectra of histone core protein. (a) 270 MHz ^{1}H NMR
spectrum, with the aromatic region vertically expanded × 8. (b) 22.6 MHz ^{13}C NMR spectrum.
Both chemical shift scales are relative to TMS at 0 ppm.

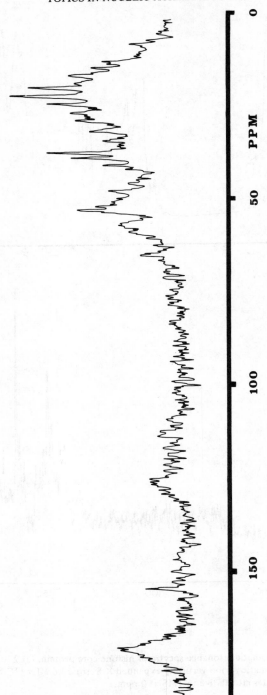

Figure 7.3 22.6 MHz ^{13}C NMR spectrum of histone core protein mildly digested by trypsin to remove the N-terminal regions of the histones.

shown that the interacting 'surfaces' show strong evolutionary conservation (Spiker and Isenberg, 1978).

In some respects, therefore, the histones appear to resemble globular enzymes. However, their primary sequences (DeLange et al., 1969a, b; Iwai, Ishikawa and Hayashi, 1970; DeLange, Hooper and Smith, 1972) suggest that each histone molecule may comprise two domains. The N-terminal sections are more basic, less evolutionarily conserved (Von Holt et al., 1979) and possess sites for in vivo modification (DeLange and Smith, 1971) whereas, the C-terminal regions are of 'normal' amino acid composition and are extraordinarily strongly conserved (Von Holt et al., 1979). Early ^{1}H NMR studies on single histone fractions (Bradbury and Crane-Robinson, 1971) together with ^{13}C NMR investigations (Lilley et al., 1975) indicated that the propensity for structure formation was lowest at the histone N-termini. Extension of these studies to include histone complexes supported this view (Lilley et al., 1976; Moss et al., 1976a; Lilley, Pardon and Richards, 1977). Perhaps the most convincing study is that of core protein, extracted and studied in 2 M sodium chloride where charge repulsion effects are screened. The ^{13}C NMR spectrum of core protein is shown in figure 7.2. The resonance linewidths are extremely narrow compared with globular proteins of similar mass indicating the presence of considerable localised flexibility in the molecule. This includes part or parts of the protein backbone since the Cα resonances (45 to 60 ppm) are narrow, and have relatively long relaxation times (Lilley et al., 1976, 1977). Evidence that the mobile regions are N-terminal is available. They are specifically cleaved by trypsin or trypsin-like proteases (Weintraub and Van Lente, 1974; Brandt, Böhm and Von Holt, 1975), resulting in a loss of narrow lines from the spectra (Lilley, 1977) (figure 7.3). Thirty-seven and 41 residues may be removed from the N-terminal of H3 and H4, respectively, without abolishing complex formation, whereas removal of C-terminal sections prevents native complexation (Böhm et al., 1977). We conclude that extracted histone complexes exist as structured, hydrophobic C-terminal globular domains in which heterologous histone–histone interactions occur, and from which basic random coil N-terminal 'tails' emerge.

Some progress has been made in determining the shape of the histone core protein in solution. Neutron scattering measurments give radii of gyration of 30.1 ± 0.3 Å, independent of concentration or source (Pardon et al., 1978). This value is closely similar to that measured for the core of the nucleosome (see table 7.2). Higher angle neutron scattering profiles show a single secondary maximum at 38 Å, from which some shape information can be derived (Pardon et al., 1978). Whilst Fourier transformation of spherically averaged intensities from solution scattering cannot be used to solve the structure, a sine transformation, analogous to the Patterson function, is possible (Porod, 1951). Thus the experimental data are used to compute a pair distribution function, the probability of existence of any vector length within the scattering macromolecule. A core protein sine transform is shown in figure 7.4. Comparison with calculations for spherical structures (Lilley et al., 1978) shows that the more hyperbolic shape reflects an axial ratio considerably different from unity, short correlations being of higher frequency at the expense of longer ones. The intercept indicates a maximum dimension of

80 ± 5 Å. Direct calculation of scattering profiles from spherical structures (Pardon *et al.*, 1978) also fails to reproduce the features of the experimental scattering curve, particularly the high intensity secondary maximum. Better agreement has been achieved by calculation of squat cylindrical structures around 20 Å height and 70 Å diameter. The agreement has been further improved by addition of lower scattering intensity domains corresponding to regions occupied by flexible histone 'tails' in a time averaged manner.

The histone core protein isolated from chromatin in 2 M sodium chloride is thus believed to be composed of a flat, disc-shaped domain comprising the C-terminal histone regions, in which the heterologous histone–histone interactions occur stabilised by hydrophobic forces, and from which the basic N-terminal histone regions extend. As later discussed, the latter sections are likely to be of considerable importance as a primary DNA binding domain. It is to be hoped that crystallisation of the histone core protein will present additional information on the structure of this important component of chromatin.

Shape and dimensions of the core particle

Early electron microscopic (EM) images of chromatin showed apparently globular nucleosomal objects separated by spacer DNA (Woodcock, 1973; Olins and Olins, 1974; Oudet, Gross-Bellard and Chambon, 1975), where the exit and entry points were on one side of the particle (Olins *et al.*, 1977*a*). The resolution of these micrographs did not, however, permit any further description of nucleosome shape. Monomeric nucleosome preparations have also been examined by EM. Analysis of such images is potentially complicated by any asymmetry present in the particle, since this may affect orientation on the grid, and hence the projection visualised. Varshavsky and Bakayev (1975) presented EM fields containing two image types. The most frequent was circular with a darker central region whilst a few images comprised parallel lines of heavy stain accumulation between which was a less stained band. The latter might represent a side elevation of an oblate object which appears circular in plan. Further evidence for an oblate geometry came from scanning transmission EM images of unfixed and unstained chromatin. Langmore and Wooley (1975) observed 134 Å diameter nucleosomes which they interpreted as flat discs 50 Å in height by studying scattered electron intensity (Wooley and Langmore, 1978).

Neutron scattering profiles from core particles also support a flattened structure. X-ray diffraction from fibres and gels of chromatin produced series of secondary maxima at 110, 55, 37, 27 and 22 Å (Wilkins, Zubay and Wilson, 1959; Pardon, Richards and Cotter, 1974) previously interpreted in terms of a regularly super-coiled structure (Pardon and Wilkins, 1972) with a pitch of 110 Å. However, maxima at 37, 27 and 22 Å and a shoulder at 55 Å have also been obtained from X-ray scattering of core particle solutions (Richards *et al.*, 1976). Although these profiles do not discriminate well between various possible structures, Harrison and Kornberg (1976) proposed a spherical nucleosome structure on their basis. Neutron scattering allows considerable restriction in the choice of models. A typical high

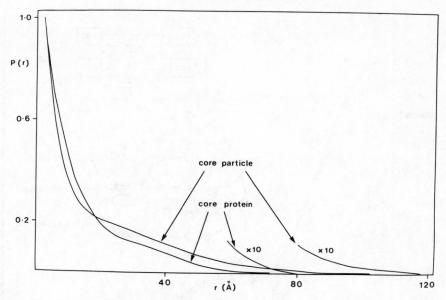

Figure 7.4   Probability distribution functions computed by sine transformation of the neutron scattering curves from core particles and core protein.

angle profile of core particles in D_2O buffer is shown in figure 7.5. The sine transform of these data is presented in figure 7.4, indicating a maximum dimension of 120 Å and, like the core protein transform, considerable deviation from spherical symmetry. Pardon *et al.* (1977) have computed scattering curves from a variety of spherical, ellipsoidal and cylindrical models of the core particle, in which the DNA and histone were constrained to concentric domains such that they gave the correct radii of gyration for the individual components and the overall assembly. They were clearly able to exclude spherical models and only one type of model (figure 7.5) was able to provide satisfactory agreement between calculated and experimental scattering profiles. This squat, cylindrical structure of height 50–55 Å and diameter 110 Å has domains of DNA (with a small amount of protein) as annuli at the top and bottom, with a flat hydrophobic histone core. It also gives good agreement with x-ray scattering profiles. Hjelm *et al.* (1977) and Suau *et al.* (1977) have studied neutron scattering from nucleosome and core particle solutions, respectively, as a function of contrast. By these means they have been able to decompose the scattered intensity into functions (Ibel and Stuhrmann, 1975) depending upon particle shape, internal scattering density fluctuations and a cross (interference) term. Whilst in theory this method increases the available data for analysis, it inevitably adds high and low contrast data resulting in poor noise levels at higher angles. The shape function was shown to support deviation from spherical symmetry, but no derivation of a unique model was made. Suau *et al.* (1977) showed that the data

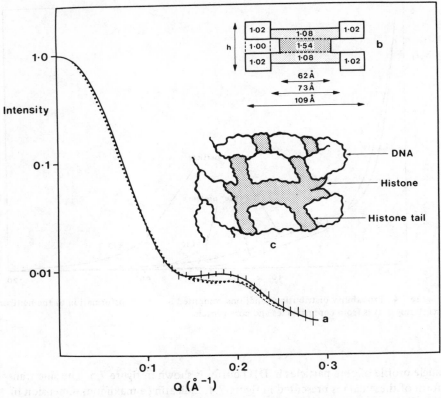

Figure 7.5 The high angle neutron scattering curve from core particles in D_2O, unbroken line. The broken lines show model calculations giving the best fit with the experimental data, corresponding to the shape and dimensions indicated. The heights (h) of the particles calculated are 50 Å (dotted line) and 55 Å (dashed line)

were consistent with models in which two turns of DNA wrapped around a central globular histone core. H1-containing nucleosome solutions have been studied by Baudy *et al.* (1976) using neutron scattering. They derived a model in which the DNA was not entirely constrained to an outer domain, but which was obtained using a questionable method of calculation and which was not in agreement with most other investigations.

An inherent limitation of scattering data is the lack of orientation, giving profiles of spherically averaged scattered intensity. Fortunately, crystals of core particles are now available (Finch *et al.*, 1977) which diffract x-rays to a resolution of about 20 Å. Crystal fragments were studied by EM from which optical diffraction allowed some phase assignment of the x-ray diffraction. Core particles were aligned in sinusoidal columns with 57 Å centre-to-centre spacing and a strong bipartite character in that direction. The unit cell contained twelve core particles, three in the asymmetric unit, each being wedge shaped, 110 Å in diameter, with greatest electron density at the top and bottom. From these data Finch *et al.* propose that the core particle contains $1\frac{3}{4}$ turns of a flat helix of DNA with a pitch of 28 Å. This is closely similar to the model derived from the neutron scattering data (Pardon *et al.*, 1977, 1978). Whilst such close agreement is gratifying, it is useful to compare the validity of each approach. Both are limited in resolution to around 20 Å, but this is deceiving since greater accuracies result from the smaller errors inherent in the radii of gyration and the unit cell dimensions. Solution scattering gives spherically averaged data which can in principle only exclude models. However, in the case of the core particle, essentially only one model was found to satisfy the data despite many calculations. Crystal data obviate this problem but, in the absence of isomorphous derivatives, require phasing from EM images of negatively stained material. Thus the resulting transform is that of the regions of the particle excluding uranyl acetate. The greatest reservation concerning the crystal structure arises from the proteolytic degradation suffered by the histones of the crystallised core particles. Similar particles have been studied in solution (Lilley and Tatchell, 1977; Whitlock and Simpson, 1977) and found to have spectroscopic, hydrodynamic and melting characteristics consistent with a slightly 'opened' structure. However, crystals are now available (Finch and Klug, unpublished data) which contain unproteolysed histone, a smaller unit cell and which diffract to a greater resolution. One important aspect of the coincidence of neutron and x-ray structures is the confirmation of the identity of solution and crystal structure.

Further evidence for a bipartite structure for the core particle comes from EM studies of particles in crystallisation buffer (Finch *et al.*, 1977). Wavy columns of particles have been observed with associated pairs of 27 Å striations. In most cases the striations were not quite parallel, giving a wedge shape related by mirror symmetry. Finch *et al.* suggest that these correspond to views at right angles to a twofold axis, as could result from the arrangement of the histone molecules about a true dyad axis. Dubochet and Noll (1978) have also made EM investigations of core particles from crystallisation buffer. They observe the formation of arcs and helices of core particles, in addition to hexagonal crystals, and have used image reconstruction techniques to improve the contrast. Like Finch *et al.*, they observe a basic wedge shaped unit 110 Å by 60 Å, and they interpret the existence of concentric arcs of these particles to imply some variability in wedge angle. Dubochet and Noll also invoke a rotation of the two halves in order to generate their observed helical assemblies of core particles which are about 1000 Å in diameter. Rows of nucleosomes are visible with a pitch angle of about 30°. The implied conformational flexibility of the nucleosome will be discussed in greater detail in the final section headed The nucleosome as a dynamic structure.

THE CONFORMATION OF DNA IN THE NUCLEOSOME

In some respects the solution of the shape and dimensions of the nucleosome core particle raises more questions than it answers. In particular we may ask how 145 bp of DNA, which would exhibit rigid rod-like behaviour in free solution, are bent into some sort of circle or helix with a diameter of around 100 Å. What are the energy requirements for this, what are the fundamental DNA–histone interactions which bring it about, how does the DNA alter its geometry to bend in this manner and what are the topological consequences of such alterations? These questions are important for two reasons. Firstly, because they help to define nucleosome structure in some detail and, secondly, because their solution may allow insight into possible disassembly mechanisms required during genetic expression.

Histone–DNA interaction

NMR evidence for a 'two-domain' model for histone complex structure has been discussed earlier. All complexes studied, including core protein in 2 M sodium chloride solution, appear to consist of structured C-terminal domains reminiscent of globular enzymes, with unstructured flexible N-terminal regions which have a high basic amino acid content. It has been suggested (Bradbury and Crane-Robinson, 1971; Lilley et al., 1977) that the N-terminal regions are of primary importance for histone–DNA interaction. If these sections exist as DNA combining domains in chromatin, then on release from DNA it might be expected that there is no residual structure. Of course, this does not exclude the C-terminal region from having a role in DNA interaction. It has been pointed out (Lilley et al., 1977) that the C-terminal basic regions also have some charged residues which are likely to form a hydrophilic surface upon the globular core. These will undoubtedly interact with the DNA, and may define a path for its folding about the core.

There is evidence that the N-termini do form distinct domains in chromatin as well as in the isolated complexes. Proteolytic enzymes like trypsin insert fairly specific cleavages in the N-terminal histone regions in chromatin, core particles and core protein (Brandt et al., 1975; Weintraub, Palter and VanLente, 1975; Weintraub and Van Lente, 1974). Thus a trypsin-like endoprotease in cycad introduces specific cleavages into histones H4 and H3 after residues 16 and 23 respectively. The fragment patterns produced by tryptic cleavage of core protein and core particles are very similar both in size and kinetics (Lilley and Tatchell, 1977) and the effects upon the physical properties of core particles (Lilley and Tatchell, 1977; Whitlock and Simpson, 1977) indicates a loosening or unfolding of the DNA. It seems, therefore, that the N-terminal domains are important for maintaining nucleosome structure but not all important since the unfolding is incomplete. Furthermore, particles resembling nucleosomes may be reconstituted from histone cores lacking their N-termini (Whitlock and Stein, 1978). The results of trypsin treatment suggest that the N-terminal regions are not just an integral part of the globular core which has artefactually unfolded on removal of DNA. Investigation by vibrational spectroscopy further precludes such an unfolding. Studies by infrared (Cotter and Lilley, 1977) and laser Raman (Thomas, Prescott and Olins, 1977) spectroscopy indicate that the core particle spectra are effectively summed core protein and DNA spectra, that is, there is no extra protein secondary structure induced upon binding of histone to DNA. Thus the N-terminal histone regions are not involved in any regular structure by combination with DNA, in pronounced contrast to the C-terminal regions which appear to be very high in α-helical structure (Thomas, Prescott and Olins, 1977).

Since the three-dimensional arrangement of the histones within the core particle is not known it is of interest to discover contact points along the nucleosomal DNA. Simpson (1976) has used a combination of chemical crosslinking and $5'[\alpha\text{-}^{32}P]$ end labelling to demonstrate that histone H3 has a contact point close to the $5'$-terminus. Using similar techniques, Mirzabekov et al. (1978) have attempted to map out many contact sites and have concluded that a single histone may have several contacts with the DNA. This result might be expected for DNA which is bound by both the N-terminal domains and the globular surface of the histone core, but which complicates the analysis, particularly in view of the relatively poor resolution available from such experiments. As well as the $5'$ end contact point for H3, histones H3 and H4 appear to occupy all the central part of the 145 bp DNA together with a H2A contact. The average lengths of the contacts are about 11 bp, that is, about one turn of helix, although many are shorter than this. Mirzabekov et al. (1977) have also shown that there is differential protection of the DNA grooves by histone. By examining protection against methylation conferred by histone binding they conclude that the minor groove (adenine N3) is well exposed whilst the major groove (guanine N7) is protected by 14 per cent. Thus both grooves are relatively open to dimethyl sulphate which could result either from the extent of contact or their mobility. In this respect it is interesting that the histone lysyl residues which must be involved in binding to DNA are readily accessible to chemical modification by imidoesters. It should not be surprising if the ionic interactions between positively charged amino acid side chains and DNA phosphate groups are quite mobile, since there is no inherent spatial dependence for this interaction. Provided the total energy surface for a number of such interactions remains reasonably flat then considerable mobility may result. Lysyl residues of another DNA binding protein, fd phage gene 5 protein, have been shown (Coleman et al., 1976) to be remarkably mobile despite DNA binding, and for the nucleosome similar mobility may confer significant functional advantages.

Supercoiling

The concept of DNA supercoiling applies to several aspects of eukaryotic chromatin structure. Supercoiling is a consequence of the plectonemic two-stranded structure of DNA, such that the strands may become interlinked to a variable extent, which is manifest when the DNA is covalently closed circular or otherwise constrained. In a closed circular DNA molecule, which may be topologically represented by a ribbon, the number of times one strand winds about the other is termed the linking number (Lk), or topological winding number, and is invariant in the absence of covalent bond fission (Vinograd and Lebowitz, 1966; Fuller, 1971). This is related to two variable parameters (Fuller, 1971), the twist (Tw), which is the number of turns about the duplex axis, and the writhing number (Wr), which is the number of turns of the duplex axis about a superhelical axis. These parameters obey the relationship

$$Lk = Tw + Wr \qquad (7.2)$$

For relaxed DNA in a plane it is self-evident that by definition:

$$Lk = Tw \qquad (7.3)$$

However, if the linking number is changed from the most stable value then either the twist, the writhing or both will alter, subject to equation 7.2. It is usually more convenient to normalise these quantities to those for relaxed DNA, in which case:

$$\Delta Lk = \Delta Tw + Wr \qquad (7.4)$$

Wr can also be termed superhelix density. From the neutron and x-ray investigations of the core particle it was concluded that the DNA makes about $1\frac{3}{4}$ turns about the histone core. Germond et al. (1975) and Keller (1975) have formed nucleosomes on SV40 DNA, relaxed the complex with nicking-closing enzyme and measured the resulting change in linking number by gel electrophoresis. Comparing this with the number of nucleosomes formed as measured by EM, they concluded that $\Delta Lk = -1.2$ per nucleosome. How can this be reconciled with the $1\frac{3}{4}$ turns estimated above? Firstly, the ΔLk measured can partition between twist and writhe; the $1\frac{3}{4}$ turns are pure writhing. It is possible that the adoption of nucleosomal conformation involves an alteration in the twist of the DNA, although it will be seen later that nuclease digestion experiments do not support this possibility. Secondly, the neutron and x-ray structures cannot totally define the path of the DNA, nor can they preclude that a small segment of DNA has a different route, for example through the nucleosome centre, which would have profound topological consequences (Crick, 1976). However, none of the biophysical studies provide any evidence for such a structure which would also be difficult to reconcile with the reversible association and dissociation of crosslinked histone core and DNA (Stein et al., 1977). A further possibility is that the difference between 1.2 and 1.75 lies in 'extra' writhing which results on the adoption of higher order coiling in the SV40 mini-chromosomes on which the change in linking is measured.

Some estimates of the energy required to deform linear DNA to the shape of the nucleosome may be made. This has been treated in some depth by Camerini-Otero and Felsenfeld (1977). Neglecting any change in twist, they estimate a requirement for 20 to 28 kcal per mole of nucleosome. The prior presence of negative superhelix density (underwound DNA) facilitates the formation of nucleosomes on closed circular DNA (Germond et al., 1975) as would be predicted since this effectively raises the free energy of the reactants. The energy required to bend the DNA can come from histone-DNA and histone-histone interaction. Camerini-Otero and Felsenfeld (1977) have considered these sources separately in order to compare bending and histone-histone interaction enthalpies, the process thus being unimolecular. Histone complex formation constants have been carefully measured by spectroscopy (D'Anna and Isenberg, 1974) from which Camerini-

Otero and Felsenfeld calculate average solution dimerisation free energies of −12 kcal per mole, which may rise to −20 kcal per mole for DNA bound dimerisation. It is clear, therefore, that histone–histone interactions may drive nucleosome formation, that is, DNA bending.

Further, it appears that histones H3 and H4 may have a special responsibility for DNA bending, since these histones alone may generate supercoils in closed circular DNA (Bina-Stein and Simpson, 1977; Camerini-Otero and Felsenfeld, 1977). The resulting 'chromatin' has many properties in common with normal chromatin (Camerini-Otero, Sollner-Webb and Felsenfeld, 1976; Bina-Stein and Simpson, 1977; Moss et al., 1977 and Oudet et al., 1978). Nucleosome-like particles have been generated containing an octamer of H3 and H4 alone, and around 140 bp of DNA (Moss et al., 1977; Simon, Camerini-Otero and Felsenfeld, 1978; Stockley and Thomas, 1979). A question thus remains on the role of the remaining histones H2A and H2B in the generation of natural chromatin.

Theoretical approaches to DNA conformation

Whilst interaction of DNA with histone has been shown to be energetically capable of bending the double stranded DNA we have not considered how the conformation of the double helix may deform in the process. Essentially there are two possible types of conformational alteration which might occur. Continuous bending may happen, in which neighbouring base pairs are all related by the same symmetry operation. Alternatively, discontinuous kinking may occur to generate short linear lengths of duplex interrupted by kinks at intervals. There is at present relatively little hard evidence with which to distinguish these possibilities, but there are a number of theoretical treatments and energy calculations.

A discontinuous kink model was first proposed by Crick and Klug (1975) in which 10 bp of B DNA were interrupted by a kink of approximately 100°. This kink model required the loss of one base-stacking interaction and the alteration of the C4′–C5′ torsion angle by 120° to an alternate staggered conformation. Whilst it is not easy to estimate the energy requirement for formation of such a kink it is probably not more than 2 kcal per mole of kink. Sobell et al. (1976) have proposed an alternative kink in which the ribose ring pucker is altered from C2′ endo to C3′ endo-(3′-5′)-C2′ endo with partial loss of base stacking. This was predicted on the basis of crystallographic studies of ethidium intercalation (see chapter 8). If such kinks are placed at 10 bp intervals along DNA, the resulting structure is a left-handed toroidal helix of 100 Å diameter, in reasonable agreement with the known nucleosome dimensions.

However, the weakest aspect of these theoretical models for DNA conformation is the lack of any reliable estimates for the energy required to produce the deformations. Levitt (1978) has attempted to rectify this situation with calculations based on empirical energy functions, and, in so doing, has shifted the tide of opinion from kinked structures to ones with continuous bending of DNA. Iterative procedures were used in which the total energy of the structures was minimised, or as nearly so as is possible for the large numbers of degrees of freedom present. Initially, the x-ray crystallographic conformation of B form DNA (Arnott and Hukins, 1973) was itself minimised, resulting in small alterations of atomic positions and torsion angles and a change of ring pucker. Interestingly, the twist of the energy 'relaxed' DNA is no longer the integral 10 bp per turn seen in crystals, where the crystal packing forces may themselves exert an influence, but rises to 10.6 bp per turn. Using the energy minimisation procedure, Levitt (1978) has shown that the energy refined DNA may be smoothly deformed into a superhelix of 90 Å diameter and 55 Å pitch, that is the dimensions of the nucleosome. Small variations in torsion angle and base tilt result, and the twist is close to 10.0 bp per turn. However, as explained below, this value for DNA twist in chromatin is not in agreement with experimentally determined values.

A very similar, smoothly bent superhelix structure has been proposed by Sussman and Trifonov (1978). They have produced an energy refined left-handed helix with 100 Å diameter and 55 Å pitch, in which there are continuous small variations of torsion angle with 10 base periodicity along the DNA. All refined bond lengths and angles are within 0.01 Å and 2.6°, respectively, of standard values and the root mean square deviation of the energy refined DNA axis from the initial superhelix axis is 0.01 Å. A more radical proposal for a smooth superhelix (Sasisekharan, Pattabiraman and Gupta, 1978) involves a new DNA structure without linkage, giving difficulties in accounting for the superhelical properties of closed circular DNA and having the wrong pitch (100 Å) to be accommodated into the nucleosome.

In summary, whilst stereochemically reasonable kinks may be described, energy refinement calculations indicate that they are not required in order to bend DNA with a 50 Å radius. However, such calculations ignore the role of the histone core, and there is no *a priori* reason why the minimum energy solution for a pure DNA superhelix should be the actual arrangement in chromatin. There is relatively little experimental data from which to distinguish the theoretical possibilities. Circular dichroism (CD) spectra of core particles show a pronounced reduction in molar ellipticity at 280 nm compared with that of pure DNA (Mandel and Fasman, 1976) which is increased to a more 'DNA-like' value by trypsin cleavage of the histone N-termini (Lilley and Tatchell, 1977; Whitlock and Simpson, 1977). Our understanding of CD processes is not sufficient to permit detailed analysis of DNA conformation in the core particle, however, and the exact origin of the change at 280 nm is not clear. No significant changes in DNA vibrational modes have been identified by laser Raman spectroscopy (Thomas *et al.*, 1977) which seems consistent with either kinked or bent structures. The environment of the phosphate groups of DNA and core particles have been studied by ^{31}P NMR (Cotter and Lilley, 1977; Kallenbach, Appleby and Bradley, 1978) which is very sensitive to torsional distortion (Gorenstein and Kar, 1975). The absence of a chemical shift difference between the resonances of DNA and core particles, shown in figure 7.6, indicates that the tetrahedral geometry of the phosphodiester bonds must be closely similar. Kallenbach *et al.* (1978) use their data to exclude kinked models, which may not necessarily be justifiable for all such structures.

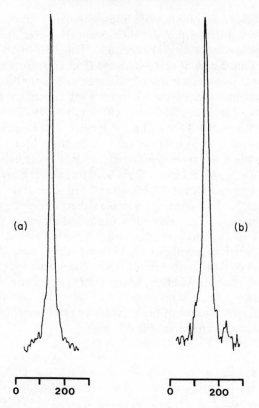

Figure 7.6   36.4 MHz ^{31}P NMR spectra of (a) 145 bp DNA and (b) 145 bp core particles in 10 mM tris, pH 7.5, 0.7 mM EDTA. The chemical shifts for both species are 121 Hz relative to TMP

Endonuclease digestion of nucleosomal DNA

Valuable information about DNA conformation may be obtained by studying the cleavage of nucleosomal DNA by endonucleases which cut *within* the nucleosome. Noll (1974*b*) first showed that pancreatic DNase I introduces single strand nicks into DNA of chromatin at sites which are multiples of around ten nucleotides. He pointed out that this could result from *B* form DNA which is bound to a protein surface, or DNA which is kinked at ten nucleotide intervals. It is interesting to note that a calcium phosphate precipitate of DNA is also cleaved at ten nucleotide intervals by DNase I (Liu and Wang, 1978), this result presumably arising from surface binding to microcrystals. This would leave the most exposed sites on the DNA having the same repeat as the pitch of the DNA helix, that is around 10 nucleotides.

Experiments of this kind cannot yield information on relative cleavage frequencies at different sites within the nucleosome. However, these data are available from experiments in which kinased $[5'-^{32}P]$ labelled core particles are digested, and the resulting gel of single-stranded DNA fragments is autoradiographed. The $5'$ label provides a frame of reference since only those fragments which possess an original nucleosome $5'$ terminus will contribute to an autoradiographic band (Simpson and Whitlock, 1976; Lutter, 1977; Noll, 1977; Sollner-Webb and Felsenfeld, 1977). There is general agreement between the results of these four groups of investigators. All ten base multiple sites are cleaved by DNase I to some extent, but there is pronounced kinetic variation between different sites. Sites at 10, 20, 40, 50 and 130 nucleotides from the $5'$ termini are cleaved relatively rapidly, whereas sites at 30, 60, 80 and 110 nucleotides are resistant. Assuming that the DNA conformation is regular, that is, smoothly bent or regularly kinked, these differential effects may come from local differences in DNA-histone environment, or from a nonintegral DNA twist, or both. Finch *et al*. (1977) have noted that sites of similar accessibility are to some extent related by translations of 80 nucleotides, and could therefore result from close spatial approximation on a superhelix with this pitch. In addition, Lutter (1978) has drawn attention to a progressive increase in rate constant from $3'$ to $5'$ along the core particle DNA. This could arise from hindered access of the nuclease by an adjacent turn of superhelix, and suggests that this helix has a left-handed sense.

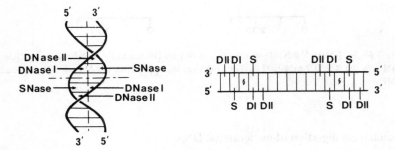

Figure 7.7   Schematic indication of the relative cutting sites for DNase I, DNase II and staphylococcal nuclease within nucleosomal DNA (see text for references)

Other endonucleases, such as DNase II and micrococcal nuclease, also cleave within the nucleosome but with different relative frequencies at the 10 nucleotide multiple sites (Whitlock, Rushizky and Simpson, 1977). Clearly the structure of the nuclease itself is another factor required in the interpretation of these experiments, although it is worth noting that even in the absence of a complete analysis, the patterns provide a useful 'fingerprint' of the native nucleosomal structure. Ten base periodicity is not restricted to endonucleases, since exonuclease III appears to undergo kinetic pauses at ten base intervals in the core particle (Riley and Weintraub, 1978).

Denaturing gels are required to observe the ten nucleotide ladders from these digestions. The reason for this is that cleavages by DNase I, DNase II and micrococcal nuclease are staggered between opposite strands (Lutter, 1977; Noll, 1977; Sollner-Webb and Felsenfeld, 1977; Sollner-Webb, Melchior and Felsenfeld, 1978), that is, the resulting fragments are non-flush ended. Experiments in which protruding single-stranded regions are removed using S1 nuclease or filled in using DNA polymerase have located these relative positions. For DNase I good agreement has been reached that cuts occur at $10n + 8$ (5'P), $10n + 2$ (3'OH) nucleotides, shown in figure 7.7. The sites for DNase II and staphylococcal nuclease are symmetrically related to those for DNase I (Sollner-Webb *et al.*, 1978) also shown in figure 7.7. These sites are consistent with a constant nuclease recognition site, where cleavage occurs at one or two nucleotides distant. Unfortunately, however, the results do not allow the hoped for distinction between bending and kinking models. Lutter (1979) has used sequencing gels to obtain higher resolution patterns of DNase I digests in which individual base multiples are resolved. By these he has calculated the DNA twist of the nucleosome to be 10.4 nucleotides. This is close to current estimates for solution DNA (Wang, 1978), but higher than the energy minimised value of Levitt (1978). Thus changes in twist between nucleosomal and free DNA cannot account for the earlier discussed apparent discrepancy between the writhe of nucleosomal DNA and the change in linkage. Trifonov and Bettecken (1979) have, however, used nonintegral pitch to explain the variation in DNase I cleavage frequency along the nucleosome DNA.

To summarise the status of DNA conformation in the nucleosome, the position is still largely one of ignorance. It is clear that the histones can provide enough energy for DNA coiling and that it is not necessary to invoke the presence of kinks in the coiling process. However, whether or not the DNA is actually bent or kinked is still not known with any certainty, and we are likely to require a crystallographic solution to this problem.

THE CHROMATIN THREAD

Having reviewed the available information on the structure of the individual nucleosome we will now consider how these are arranged to form the nucleofilament thread. The arrangement and extension of the linker DNA between core particles will be considered, and the relative arrangement of neighbouring nucleosomes. It is generally expected that regular or semiregular higher order structures will be formed, and some kind of organisation must extend up to the level of the whole chromosome. A major problem becomes the likely fragility of the structures. It has been well established that shear forces may disrupt nucleosomal structure (Noll, Thomas and Kornberg, 1975), and even greater care is probably essential in order to preserve higher order structure. Whilst the physical methods are available for the study of large assemblies of nucleosomes it is probably the biochemical technique which is rate limiting for progress in this area.

Separation and orientation of nucleosomes

Whilst EM images of chromatin as globular nucleosomes well separated from their neighbours by an extended DNA spacer (Woodcock, 1973; Olins and Olins, 1974; Langmore and Wooley, 1975; Oudet *et al*., 1975) were extremely helpful in the conceptual formation of contemporary chromatin structure, we may ask if this is the true state of the chromatin thread *in situ*. It is known that the Millar spreading technique (Millar and Beatty, 1969) applies stretching forces which could artificially extend the structure, and EM pictures of chromatin prepared by alternative procedures frequently have a more continuous appearance (Ris and Kubai, 1970; Pooley, Pardon and Richards, 1974). The relative extension visualised appears to be influenced by the presence of lysine-rich histone H1 causing an apparent contraction of the chromatin (Thoma and Koller, 1977; Griffith and Christiansen, 1978; Keller *et al*., 1978).

In recent studies of nucleosome spacing along the chromatin thread nuclease digested chromatin has been used in order to avoid shear forces, and hence artificial extension, during isolation procedures (Noll *et al*., 1975). Such digests may be fractionated into discrete sized chromatin oligomers (for example, see figure 7.1) for which radius of gyration measurements allow estimation of spacer extension, and longer chromatin, for which mass per unit length and cross sectional radius of gyration measurements give an indication of the distribution of nucleosomes along the fibre.

Early x-ray scattering measurements upon chromatin prepared without nuclease digestion (Luzzati and Nicolaieff, 1963; Bram and Ris, 1971) gave mass per unit lengths considerably smaller than a later study (Sperling and Tardieu, 1976) of nuclease prepared chromatin. Whilst the former measurements were consistent only with extended structures the latter investigation, of chromatin with an average length of about 40 nucleosomes, gave a cross sectional radius of gyration of 38 Å and a mass per unit length of 1240 electrons per Å. This corresponds to a 110 Å diameter fibre with nucleosome centres separated by 100 Å. Bram *et al*. (1977) have used neutron scattering in a similar investigation to extend data collection to lower angles, and have obtained still larger values for these parameters. In a more recent investigation at higher angles Baudy and Bram (1978) have obtained mass per unit length values comparable to those of Sperling and Tardieu (1976).

Neutron scattering has been used to measure radii of gyration for di, tri, tetra and hexanucleosome derived from chicken erythrocyte chromatin (Cotter *et al*., 1979). The values obtained are not consistent with fully extended spacer DNA sections, and are reduced by the presence of histone H1. Thus for the dinucleosome the radius of gyration of 72 Å (corresponding to a centre-to-centre nucleosome spacing of 120 Å) is reduced to 51 Å (centre-to-centre spacing 66 Å) when H1 is present. In the preparation of H1 depleted chromatin, involving exposure to relatively high ionic strengths, it appears that histone cores may 'slide' together, and a small proportion of 'spacerless' oligomers which are multiples of around 145 bp are produced in the ensuing nuclease digestion (Klevan and Crothers, 1977; Tatchell and Van Holde, 1978). Neutron scattering measurements upon such a

280 bp dimer have given a radius of gyration of around 50 Å, closely comparable with a normal length dimer containing H1 (Cotter *et al.*, 1979). Tatchell and Van Holde (1978) interpret DNase I digestion results upon these species in terms of a continuous DNA superhelix wrapped around stacked protein cores. Lohr, Tatchell and Van Holde (1977) have also observed extended DNase I ladders, particularly evident for yeast chromatin, from which they suggest that regions of similar closely packed cores may occur *in situ*. These results may, however, reflect inter-core interaction and spacer organisation. Histone H1 clearly interacts primarily with the spacer DNA, since this histone is lost from nucleosomes as nucleases cut into the spacer (Varshavsky, Bakayev and Georgiev, 1976; Noll and Kornberg, 1977; Todd and Garrard, 1977), but how the contraction is achieved is not understood. It is entirely possible that H1 interacts both with the spacer and the core particle.

Whilst it seems likely that the histone cores are not separated by extended spacers in native chromatin, we cannot be sure of how the cores are arranged relative to the fibre axis. Individual core particles show a tendency to stack with their cylindrical axes parallel to the fibre axis (Finch *et al.*, 1977; Dubochet and Noll, 1978), but this cannot be assumed to be true for connected nucleosomes with H1 present. As an alternative arrangement, core particles may associate edge-to-edge, thereby giving an approximately 110 Å centre-to-centre spacing. The origin of the 110 Å reflection obtained by x-ray diffraction from oriented chromatin gels (Pardon *et al.*, 1974) is not clear, and the reader is recommended to read Pardon and Richards (1979) for an extensive discussion of these data. Contrast variation studies of neutron diffraction maxima obtained from high concentration chromatin fibres (Baldwin *et al.*, 1975; Pardon *et al.*, 1978) indicate that the intensity of the low angle maximum is zero at 10 per cent D_2O in the buffer. This indicates that there is a spatial separation of hydrated DNA and relatively hydrophobic protein along the fibre axis. From the known organisation of the core particle this is most readily accommodated by a model in which the particles stack with coincident axes. However, this cannot be regarded as conclusive evidence and is further complicated by the possible existence of higher levels of organisation.

Higher order structure

Up to this point we have considered the organisation of nucleosomes along the axis joining their collective centres, which could be linear for the purposes of that discussion. The term higher order structure is taken to mean the path of this axis resulting in possible interactions between nucleosomes which are not adjacent neighbours on the DNA. Models have been proposed in which the axis is coiled to form a regular superhelix or solenoid of chromatin, or constrained to form discrete globular assemblies of nucleosomes, sometimes called superbeads.

Under some conditions chromatin has been visualised in the EM as a fibre thicker than can be accounted for by a simple thread of nucleosomes (Pooley *et al.*, 1974; Oudet *et al.*, 1975; Finch and Klug, 1976; Olins, 1978). In some instances these 300 Å diameter fibres show some internal structure. Finch and Klug (1975) examined negatively stained chromatin applied to carbon coated grids by EM as a function of divalent cation concentration. In 0.2 mM EDTA they observed 100 Å diameter fibres, which increased to 300 Å in the presence of 0.2 mM magnesium chloride. In some micrographs striations were visible across the fibres with 120 Å spacing. On the basis of these studies, together with earlier diffraction (Pardon et al., 1974) and EM (Davies and Haynes, 1975) data, they proposed that in the presence of magnesium and histone H1, chromatin adopts a solenoidal structure in which the continuous thread of nucleosomes describes a helix of 110 Å pitch and 300 Å diameter, thus having about six nucleosomes per turn. A similar model was proposed by Carpenter et al. (1976) on the basis of an observed 9° off-meridional orientation of the low-angle neutron-diffraction maximum from chromatin fibres.

Solution scattering experiments also support helical models similar to these. Campbell, Cotter and Pardon (1978) have measured radii of gyration of defined chromatin lengths which are most consistent with flexible helical structures of approximately fixed diameter but variable pitch, such that with increasing ionic strength the pitch is reduced until the structure becomes coil bound. Knobbly fibres of about 150 Å diameter have been visualised by EM (Tsanev and Petrov, 1976) from rat liver chromatin, which could arise from open helical structures, or a distribution of nucleosomes about a central screw axis. Shaw and Schmitz (1976) have proposed a further flexible coil model on the basis of light scattering studies of chromatin at low ionic strength.

Discontinuous higher order structures have been proposed in other laboratories. Renz et al. (1977) and Stratling, Muller and Zentgraf (1978) have presented micrographs of 200 Å diameter superbeads spaced along chromatin which require both Hl and relatively high ionic strength to be visualised. Renz et al. (1977) have observed a cooperative H1-dependent hydrodynamic transition as a function of ionic strength, interpreted in terms of superbead formation. Stratling et al. (1978) have used mild nuclease digestion to prepare 335 assemblies containing about 16 nucleosomes, and similar structures have been prepared by Butt, Jump and Smulson (1979). It is entirely possible that superbeads and helical structures may coexist, possibly in an interconvertible situation. In some micrographs both types of assembly are visible (Olins, 1978) where each have the internal appearance of close-packed arrangements of nucleosomes.

Two points of reasonable concensus appear to emerge from these results. Firstly, chromatin seems to possess the capacity for further levels of coiling, be it continuous or discontinuous. Secondly, H1 is likely to be involved in the adoption of these structures. However, caution is essential in deriving any further general conclusions. Whilst many of these studies tell us how chromatin can fold, it does not follow that these are necessarily in vivo structures. The exact role of H1 is difficult to determine beyond a general compaction. X-ray fibre diffraction from H1-depleted chromatin is not significantly different from that from H1-containing specimens (Richards and Pardon, 1970; Bradbury et al., 1972), thereby precluding any further analysis by this technique.

A possible probe for higher order structure in the nucleus has emerged from studies by Altenburger, Hörz and Zachau (1976) of DNase II digestion as a function of ionic strength. In the absence of either H1 or divalent cations this enzyme cleaves chromatin into the normal multiples of 200 bp. However, if both H1 and divalent cations are present a new pattern is generated with multiples of 100 bp. These authors propose that in the presence of these effectors, a higher order structure is generated such that new sites at the nucleosome centres are presented for nucleolysis by DNase II. Of course the experiments give no clue as to the nature of this structure but they do suggest the presence of some type of further coiling within the nucleus.

Chromosomes

Some gross ultrastructural characteristics of whole chromosomes have been appreciated for a long time. Whole mount electron micrographs of chromosomes (Dupraw, 1970) indicate that they consist of discrete elongated structures with a basic fibrular composition, and having distinct functional domains such as the centromere and the telomere. There is a considerable need, however, for more information to bridge the gap between the structure at the level of the chromatin fibre and at that of the whole chromosome.

Scanning EM images of whole metaphase chromosomes (Sedat and Manuelidis, 1978; Wray et al., 1978) show compact structures with uneven knobbly surfaces, and 2000 Å tubes in parallel or orthogonal arrays have been observed by high voltage EM studies of nuclei (Sedat and Manuelidis, 1978). These thicker fibres could represent a yet further stage of coiling of the chromatin thread. An alternative hollow tube arrangement for the mitotic chromosome has been proposed by Bak, Zeuthen and Crick (1977), although the evidence for such structures cannot be described as convincing.

An important aspect of whole chromosome organisation was uncovered by the discovery that protein-depleted whole chromosomes have sedimentation properties which alter as a function of ethidium bromide concentration in a biphasic manner. The observations of Cook and Brazell (1975) imply that the DNA of the chromosome has linkage, that is, points of topological constraint, such that it has potential superhelix density. By a study of target size for γ-radiation damage they were able to estimate sizes of 1500 Kb for the superhelical units. Equivalent observations were subsequently made by Benyajati and Worcel (1976) but these authors estimate considerably smaller loop sizes of about 85 Kb. Igo-Kemenes and Zachau (1978) have studied the kinetics of fragment release from chromatin upon digestion by nucleases, and thereby estimated that the DNA exists as domains or loops with an average length of 34 Kb. These observations are strongly reminiscent of the folded chromosome of Escherichia coli (Worcel and Burgi, 1972) and similar structures have been identified in lower eukaryotes (Pinon and Salts, 1977).

Perhaps the most convincing demonstration of chromosome loops comes from the work of Laemmli et al. (Paulson and Laemmli, 1977). They have presented micrographs of histone-depleted metaphase HeLa chromosomes which show a central core (or 'scaffold') of similar shape to the intact chromosome, from which DNA loops between 30 and 90 Kb emanate. Thus there is good agreement on the sizes of chromosomal loops, with the exception of the estimates of Cook and Brazell (1975) which could be in error since their sizing method is somewhat indirect. Paulson and Laemmli (1977) observe that the loop ends occur very close together at the core. Using nucleases Adolph *et al.* (1977) have isolated a purified scaffold which has a uniform width of 0.4 μm in fixed specimens. About 30 scaffold proteins have been separated on SDS gels and Campbell *et al.* (1979) have obtained antisera to some non-histone proteins found on these central regions.

In summary, many aspects of chromosome folding are unclear and require the acquisition of hard data. It seems highly probable that the basic thread of nucleosomes is further coiled once or several more times to form a reasonably organised structure for the chromosome. The entire length of DNA in a single chromosome is segmented into domains or loops, which may have important functional significance, and which may be related to the banding of chromosomes by stains and to banding and puffing phenomena in polytene chromosomes.

THE NUCLEOSOME AS A DYNAMIC STRUCTURE

In reviewing nucleosome and chromatin structure we have considered them only in a static sense up to this point. Most macromolecular assemblies are far from static at normal temperature, this being particularly evident for lipids and membranes (Levine *et al.*, 1972; Lakowicz and Weber, 1973). Motion in molecules covers a wide frequency spectrum from rotational motions of amino acid side chains on the surface of a globular protein, which could be on a time scale in the 10^{-9} s region, down to much slower motions involving large macromolecular domains, where time scales of around 10^{-4} s might be expected. In addition to such motions it is possible to have switching between two conformations in response to specific ligand binding. It is predictable that such facets of molecular dynamics will apply to the nucleosome. In addition to these considerations, there are good biological reasons for proposing that the nucleosome structure may be perturbable. Under many circumstances chromosomes are metabolically highly active, and the DNA must be accessible to the enzymes responsible for RNA and DNA synthesis, repair, recombination and so on. Since it seems unlikely that DNA strand separation will be possible in the highly condensed state of a folded chromatin thread we may consider how its structure may be altered to facilitate the passage of nucleic acid polymerases.

Dynamic nucleosomes *in vitro*

It was noted earlier in the section on The Confirmation of DNA in the Nucleosome that ionic interactions between positively charged amino acid side chains and DNA phosphate groups may well have considerable mobility. Thus the histone N-terminal sections may effectively 'float' in an overall charge cloud of the DNA, which could explain the relative ease of access of both DNA and protein to reagents which give chemical modification of these components (Mirzabekov *et al.*, 1977; Stein, Bina-Stein and Simpson, 1977). N-terminal mobility has been observed in NMR studies of nucleosome core particles (Cary, Moss and Bradbury, 1978). Such mobility could have important consequences. Firstly, histone-DNA strand association would acquire greater fluidity such that under some circumstances the histone may be preferentially associated with just one DNA strand. Secondly, the activation energy for translocation of histone cores along DNA might be reduced. By rotating the histone core about the DNA superhelix axis the core would effectively 'screw' along the DNA. Nucleosome sliding has been observed under some circumstances, resulting in movement of nucleosomes onto previously naked DNA ligated onto SV40 chromatin (Beard, 1978) and protection of restriction targets (Steinmetz, Streek and Zachau, 1975).

Conformational transitions have been observed in nucleosomes in response to changes in ionic strength, pH and concentrations of urea and organic solvents. Fluorescent adducts of core particles have been used to provide a spectroscopic probe of nucleosome structure during changes of ionic strength and urea concentration (Dieterich, Axel and Cantor, 1978; Zama et al., 1978a). In particular, a pyrene adduct in which stacking of pyrene rings attached to different H3 single cysteine residues results in enhanced fluorescence at 460 nm has been used (Zama et al., 1978a) to observe structural transitions which alter the relative spatial arrangements of H3 molecules. Five different states have been observed as a function of ionic strength. At very low ionic strength Oudet et al. (1978) have observed the presence of around 40 paired 'half-nucleosomes' on SV40 minichromosomes by EM, and an equivalent octamer–tetramer equilibrium has been proposed on the basis of hydrodynamic studies of core protein (Chung et al., 1978). These observations are consistent with the proposed flexibility (Dubochet and Noll, 1978) of the core particle seen by EM. Using zero-length crosslinking reagents Martinson, True and Burch (1979) have shown that H2B-H4 contacts are disrupted at low ionic strength. Olins et al. (1977b) have observed two transitions in core particle structure with increasing concentrations of urea, interpreting their results in terms of an initial change in DNA conformation followed by changes in core structure. However, it should be noted that chromatin in urea continues to be digested by micrococcal nuclease in a manner suggesting differential protection of 200 bp DNA lengths by histones (Jackson and Chalkley, 1975). Thus perturbed nucleosomal structures still present DNA with the same protection and hence nuclease digestion patterns of this kind are not necessarily indicative of nonperturbed nucleosome structures. Several cooperative unfolding transitions occur as the temperature of core particle solutions is raised (Bryan et al., 1978; Seligy and Poon, 1978; Weischet et al., 1978) and core particle structure has been altered by changes of pH (Zama et al., 1978b). Changes in chromatin structure as a function of pH have been also observed by fibre diffraction studies (Staynov et al., 1979). Both studies indicate that a transition occurs around pH 5. Below this pH, diffraction maxima at 110, 55 and 37 Å are reversibly lost and a new maximum at 27 Å becomes visible. Staynov et al. interpret these results in terms of loss of the transform of the intact nucleosome structure, and subsequent close packing of elongated histone-bound DNA structures. They suggest that salt bridges between lysyl or guanido and carboxylate side chains may be important for the integrity of

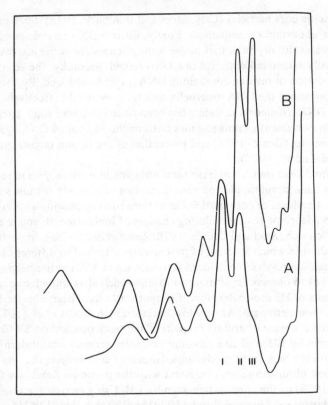

Figure 7.8   Autoradiograph scans from single-strand DNA gels of DNase I fragments from digestion of A: 160 bp nucleosomes and B: 160 bp nucleosomes preincubated with a eukaryotic RNA polymerase II. Note the increase in frequence of cleavage (III) at 70 nucleotides.

the structure. This may be true, and lysine-glutamate interactions have been suggested to be important in core protein structure on the basis of NMR data (Lilley *et al.*, 1977), but observed pH maxima are not a reliable guide for such processes since pK values can be markedly perturbed by local environmental effects, including hydrophobic character. However, it seems reasonably clear that structural changes can be induced to occur in the nucleosome by a variety of perturbants. This could indicate that the potential for structural change is present, although direct extrapolation to *in vivo* mechanisms is unwise when such major structural perturbants as urea are considered. However, similar nucleosome deforming ability seems to be possessed by a eukaryotic RNA polymerase II.On binding wheat RNA polymerase II to [5'-^{32}P] labelled 160 bp nucleosomes a conformational transition has been observed (Lilley, Jacobs and Houghton, 1979) such that a new strong site for DNase I cleavage is created close to the nucleosome centre (figure 7.8). This change has a certain biological logic and involves mild perturbation only.

Dynamic processes *in vivo*

How far can the studies of chromatin structure be applied to those parts of the genome which are genetically active? It is difficult to estimate what proportion of the cellular genetic material this might constitute, but it is probably small enough to escape detection in structural studies, particularly in the case of relatively dormant nuclei such as the terminally differentiated avian erythrocytes. Nevertheless, some genetic processes, such as replication, apply to the entire genome, and hence 'bulk' chromatin structure will be an important factor. These aspects are reviewed in greater depth in Lilley and Pardon (1979).

Nucleosomes can be released from chromatin which contain DNA sequences complementary to polysomal message (Lacy and Axel, 1975; Kuo, Saharasrabuddhe and Saunders, 1976). This implies that even active genes may have nucleosomes or structures sufficiently like them to revert to nucleosomal structure on nuclease digestion. Nucleosome-like structures have also been visualised by EM on active genes between nascent RNP strands, suggesting that elongating polymerase molecules can pass them (Foe, Wilkinson and Laird, 1976). Some structural transition seems to be a prerequisite on logical grounds, since strand separation is not possible for DNA which is surface bound to a protein core, and mechanisms have been proposed (Weintraub, Worcel and Alberts, 1976; Richards *et al.*, 1977). Bead density is relatively low on active genes and it is possible that more histone is present, to give the normal histone/DNA ratio, but in a non-nucleosomal conformation such that no 'bead' is seen in the EM. Bead density appears to correlate inversely with gene activity (Lamb and Daneholt, 1979), and no beads at all can be seen on the highly active ribosomal genes (Foe *et al.*, 1976; Franke *et al.*, 1976) and amphibian lampbrush loop genes (Franke *et al.*, 1976). EM images of replication bubbles also show beaded morphology, even very close to the replication fork (McKnight and Millar, 1977), suggesting that DNA polymerases may similarly be able to pass 'through' nucleosomes.

There is good evidence, however, that active genes and nascent chromatin do differ from 'bulk' chromatin in conformation. Active genes are more susceptible to attack by nucleases which cut within the nucleosome (Garel and Axel, 1976; Weintraub and Groudine, 1976), as is newly synthesised chromatin (Seale, 1975; Burgoyne, Mobbs and Marshall, 1976; Hildebrand and Walters, 1976; Levy and Jacob, 1978; Murphy, Wallace and Bonner, 1978; Seale, 1978; Worcel, Han and Wong, 1978). The exact level of structure which confers the DNase I sensitivity is not clear; possibly at the single nucleosome level (Weintraub and Groudine, 1976) but more likely at a higher level of organisation (Garel and Axel, 1976). Furthermore the sensitivity seems to be an all or none event, with no variation with gene activity (Garel, Zolan and Axel, 1977) and remaining after the gene product is no longer being synthesised (Weintraub and Groudine, 1976; Palmiter *et al.*, 1978). The conformational change is, however, rather precise in terms of the region of chromatin so affected (Flint and Weintraub, 1977). Nascent chromatin is also hypersensitive to nucleolysis (Seale, 1975; Burgoyne, Mobbs and Marshall, 1976; Hildebrand and Walters, 1976; Levy and Jacob, 1978; Murphy, Wallace and Bonner,

1978; Seale, 1978; Worcel, Han and Wong, 1978) and appears to be closer packed than normal chromatin (Levy and Jacob, 1978; Seale, 1978). A maturation process occurs over a time scale of a few minutes during which these characteristics are lost. The adoption of normal spacer length is interesting, suggesting that nucleosome sliding can occur *in vivo*, at least during the replication process. Newly synthesised histone is deposited as an octamer, and segregates through subsequent generations as an octamer (Leffak, Grainger and Weintraub, 1977). This implies that twofold symmetry can be temporarily lost at the replication fork and is evidence for relatively mobile DNA–histone interactions *in vivo*.

Further description of the activated structure of chromatin is desirable. Fractionation of chromatin in order to purify transcribed sections has been partially successful (Berkowitz and Doty, 1975; Gottesfeld, Murphy and Bonner, 1975; Bloom and Anderson, 1978; Levy-W and Dixon, 1978a; Tata and Baker, 1978). Using a variety of methods, mainly based upon selective nuclease digestion, purifications of around fivefold have been achieved. Although hard numbers are unavailable it seems unlikely that 20 per cent of total chromatin is in an active conformation, in which case these preparations are probably impure. Some structural changes have been observed, including hydrodynamic properties (Gottesfeld and Butler, 1977), enrichment in RNA (Gottesfeld and Butler, 1977; Hendrick, Tolstochev and Randlett, 1977), non-histone proteins (Gottesfeld et al., 1975; Lau and Ruddon, 1977; Levy-W and Dixon, 1978a) and covalently modified histones (Davie and Candido, 1978) and enhanced nuclease accessibility (Berkowitz and Doty, 1975; Gottesfeld and Butler, 1977). Some authors observe selective release of certain non-histone proteins on DNase I digestion of active genes in chromatin (Vidali, Boffa and Allfrey, 1977; Levy-W and Dixon, 1978b; Mayfield et al., 1978; Defer et al., 1979) but this has been disputed for thymus and liver nuclei (Goodwin and Johns, 1978; Billett, 1979). Active genes have also been proposed to be hyperacetylated (De Lange et al., 1969b). Hyperacetylation of bulk chromatin is possible in tissue culture (Riggs et al., 1977) and the resulting chromatin is DNase I sensitive (Sealy and Chalkley, 1978; Simpson, 1978; Vidali et al., 1978) in a manner strongly reminiscent of both active genes (Garel and Axel, 1976; Weintraub and Groudine, 1976) and proteolysed chromatin (Lilley and Tatchell, 1977; Whitlock and Simpson, 1977). Very recently Weisbrod and Weintraub (1979) have presented evidence to indicate that HMG proteins 14 and 17 may be responsible for the DNase I sensitivity of active genes. It is probable that structural investigation of chromatin activation will require further advances in purification technique. Many of the problems inherent in the preservation of higher order structure will be even greater with active chromatin. At present there are several candidates for activation mechanisms, including hyperacetylation, non-histone protein binding and altered higher order structure. It is likely that several effects are required to act in concert and the combination may be subtle and difficult to elucidate. The long term goal of understanding how a eukaryotic gene operates at the nucleoprotein level is, however, both fascinating and important.

ACKNOWLEDGEMENTS

The author thanks Drs B. M. Richards and J. F. Pardon for generating his interest in eukaryotic chromatin, and for their continuing stimulation. He also thanks Dr J. F. Pardon for comments during the preparation of this chapter.

REFERENCES

Adolph, K. W., Cheng, S. M. and Laemmli, U. K. (1977). *Cell,* 12, 805.
Altenburger, W., Hörz, W. and Zachau, H. G. (1976). *Nature, Lond.,* 264, 517.
Arnott, S. and Hukins, D. W. L. (1973). *J. molec. Biol.,* 81, 93.
Bak, A. L., Zeuthen, J. and Crick, F. H. C. (1977). *Proc. natn. Acad. Sci., U.S.A.,* 74, 1595.
Baldwin, J. P., Boseley, P. G. and Bradbury, E. M. (1975). *Nature, Lond.,* 253, 245.
Baudy, P. and Bram, S. (1978). *Nucl. Acids Res.,* 5, 3697.
Baudy, P., Bram, S., Vastel, D. and Lepault, J. (1976). *Biochem. biophys. Res. Commun.,* 72, 176.
Beard, P. (1978). *Cell,* 15, 955.
Benyajati, C. and Worcel, A. (1976). *Cell,* 9, 393.
Berkowitz, E. M. and Doty, P. (1975). *Proc. natn. Acad. Sci. U.S.A.,* 72, 3328.
Billett, M. A. (1979). *Biochem. Soc. Trans.,* 7, 381.
Bina-Stein, M. and Simpson, R. T. (1977). *Cell,* 11, 609.
Bloom, K. S. and Anderson, J. N. (1978). *Cell,* 15, 141.
Böhm, L., Hayashi, H., Cary, P. D., Moss, T., Crane-Robinson, C. and Bradbury, E. M. (1977). *Eur. J. Biochem.,* 77, 487.
Bradbury, E. M. and Crane-Robinson, C. (1971). In *Histones and Nucleohistones* (ed. D. Phillips), Plenum Press, New York, p. 85.
Bradbury, E. M., Molgaard, H. V., Stephens, R. M., Bolund, L. A. and Johns, E. M. (1972). *Eur. J. Biochem.,* 31, 474.
Bram, S., Baudy, P., Lepault, J. and Hermann, D. (1977). *Nucl. Acids Res.,* 4, 2275.
Bram, S. and Ris, H. (1971). *J. molec. Biol.,* 55, 325.
Brandt, W. F., Böhm, L. and Von Holt, C. (1975). *FEBS Lett.,* 51, 88.
Bryan, P. N., Wright, E. B., Hsie, M. H., Olins, A. L. and Olins, D. E. (1978). *Nucl. Acids Res.,* 5, 3603.
Burgoyne, L. A., Mobbs, J. D. and Marshall, A. J. (1976). *Nucl. Acids Res.,* 3, 3293.
Butt, T. R., Jump, D. B. and Smulson, M. E. (1979). *Proc. natn. Acad. Sci. U.S.A.,* 76, 1628.
Camerini-Otero, R. D. and Felsenfeld, G. (1977). *Nucl. Acids Res.,* 4, 1159.
Camerini-Otero, R. D., Sollner-Webb, B. and Felsenfeld, G. (1976). *Cell,* 8, 333.
Campbell, A. M., Briggs, R. C., Bird, R. E. and Hnilica, L. S. (1979). *Nucl. Acids Res.,* 6, 205.
Campbell, A. M. and Cotter, R. I. (1976). *FEBS Lett.,* 70, 209.
Campbell, A. M., Cotter, R. I. and Pardon, J. F. (1978). *Nucl. Acids Res.,* 5, 1571.
Carpenter, B. G., Baldwin, J. P., Bradbury, E. M. and Ibel, K. (1976). *Nucl. Acids Res.,* 3, 1739.
Cary, P. D., Moss, T. and Bradbury, E. M. (1978). *Eur. J. Biochem.,* 89, 475.
Chambon, P. (1978). *Cold Spring Harb. Symp. quant. Biol.,* 42, 1209.
Cheah, K. S. E. and Osborne, D. J. (1977). *Biochem. J.,* 163, 141.
Chung, S.-Y., Hill, W. E. and Doty, P. (1978). *Proc. natn. Acad. Sci. U.S.A.,* 75, 1680.

Coleman, J. E., Anderson, R. A., Ratcliffe, R. G. and Armitage, I. M. (1976). *Biochemistry*, **15**, 5419.
Compton, J. L., Bellard, M. and Chambon, P. (1976). *Proc. natn. Acad. Sci. U.S.A.*, **73**, 4382.
Cook, I. and Brazell, P. (1975). *J. Cell Sci.*, **19**, 261.
Cotter, R. I. and Lilley, D. M. J. (1977). *FEBS Lett.*, **82**, 63.
Cotter, R. I., Pardon, J. F., Lilley, D. M. J. and Worcester, D. L. Submitted for publication.
Crick, F. H. C. (1976). *Proc. natn. Acad. Sci. U.S.A.*, **73**, 2639.
Crick, F. H. C. and Klug, A. (1975). *Nature, Lond.*, **255**, 530.
D'Anna, J. A. and Isenberg, I. (1974). *Biochemistry*, **13**, 4992.
Davie, J. R. and Candido, E. P. M. (1978). *Cell*, **75**, 3574.
Davies, H. G. and Haynes, M. E. (1975). *J. Cell Sci.*, **17**, 263.
Defer, J., Crepin, M., Terrioux, C., Kruh, J. and Gros, J. (1979). *Nucl. Acids Res.*, **6**, 953.
De Lange, R. J., Fambrough, D. M., Smith, E. L. and Bonner, J. (1969*a*). *J. Biol. Chem.*, **244**, 319.
De Lange, R. J., Fambrough, D. M., Smith, E. L. and Bonner, J. (1969*b*). *J. Biol. Chem.*, **244**, 5669.
De Lange, R. J., Hooper, J. A. and Smith, E. L. (1972). *Proc. natn. Acad. Sci. U.S.A.*, **69**, 882.
De Lange, R. J. and Smith, E. L. (1971). *A. Rev. Biochem.*, **40**, 279.
Dieterich, A. E., Axel, R. and Cantor, C. R. (1978). *Cold Spring Harb. Symp. quant. Biol.*, **42**, 199.
Dubochet, J. and Noll, M. (1978). *Science*, **202**, 280.
Dupraw, E. J. (1970). *DNA and Chromosomes*, Holt, Rinehart and Winston, New York.
Eickbush, T. H. and Moudrianakis, E. N. (1978). *Biochemistry*, **17**, 4955.
Felsenfeld, G. (1978). *Nature, Lond.*, **257**, 177.
Finch, J. T. and Klug, A. (1976). *Proc. natn. Acad. Sci. U.S.A.*, **73**, 1897.
Finch, J. T., Lutter, L. C., Rhodes, D., Brown, R. S., Rushton, B., Levitt, M. and Klug, A. (1977). *Nature, Lond.*, **269**, 29.
Flint, S. J. and Weintraub, H. M. (1977). *Cell*, **12**, 783.
Foe, V. E., Wilkinson, L. E. and Laird, C. D. (1976). *Cell*, **9**, 131.
Franke, W. W., Scheer, U., Trendelenburg, M. F., Spring, H. and Zentgraf, H. (1976). *Cytobiologie*, **13**, 401.
Fuller, F. B. (1971). *Proc. natn. Acad. Sci. U.S.A.*, **68**, 815.
Garel, A. and Axel, R. (1976). *Proc. natn. Acad. Sci. U.S.A.*, **73**, 3966.
Garel, A. and Axel, R. (1978). *Cold Spring Harb. Symp. quant. Biol.*, **42**, 701.
Garel, A., Zolan, M. and Axel, R. (1977). *Proc. natn. Acad. Sci. U.S.A.*, **74**, 4867.
Germond, J. E., Hirt, B., Gross-Bellard, M. and Chambon, P. (1975). *Proc. natn. Acad. Sci. U.S.A.*, **72**, 1843.
Goodwin, G. H. and Johns, E. W. (1978). *Biochim. biophys. Acta*, **519**, 279.
Gorenstein, D. G. and Kar, D. (1975). *Biochem. biophys. Res. Commun.*, **65**, 1073.
Gottesfeld, J. M. and Butler, P. J. G. (1977). *Nucl. Acids Res.*, **4**, 3155.
Gottesfeld, J. M. and Melton, D. A. (1978). *Nature, Lond.*, **273**, 317.
Gottesfeld, J. M., Murphy, R. F. and Bonner, J. (1975). *Proc. natn. Acad. Sci. U.S.A.*, **72**, 4404.
Griffith, J. D. and Christiansen, G. (1978). *Cold Spring Harb. Symp. quant. Biol.*, **42**, 215.
Guinier, A. and Fournet, G. (1955). In *Small-angle Scattering of X-rays*, John Wiley, New York.
Harrison, S. C. and Kornberg, R. D. (1976). *ICN Conference on Molecular Biology*, p. 7.

Hendrick, D., Tolstochev, P. and Randlett, D. (1977). *Gene*, **2**, 147.
Hewish, D. R. and Burgoyne, L. A. (1973). *Biochem. biophys. Res. Commun.*, **52**, 504.
Hildebrand, C. E. and Walters, R. A. (1976). *Biochem. biophys. Res. Commun.*, **73**, 157.
Hjelm, R. P., Kneale, G. G., Suau, P., Baldwin, J. P. and Bradbury, E. M. (1977). *Cell*, **10**, 139.
Ibel, K. and Stuhrmann, H. B. (1975). *J. molec. Biol.*, **93**, 255.
Igo-Kemenes, T. and Zachau, H. G. (1978). *Cold Spring Harb. Symp. quant. Biol.*, **42**, 109.
Iwai, K., Ishikawa, K. and Hayashi, H. (1970). *Nature, Lond.*, **226**, 1056.
Jackson, V. and Chalkley, R. (1975). *Biochem. biophys. Res. commun.*, **67**, 1391.
Johns, E. W., Goodwin, G. H., Walker, J. M. and Sanders, C. (1975). In *The Structure and Function of Chromatin, Ciba Foundation Symposium*, vol. 28, p. 95.
Kallenbach, N. R., Appleby, D. W. and Bradley, C. H. (1978). *Nature, Lond.*, **272**, 134.
Keller, W. (1975). *Proc. natn. Acad. Sci. U.S.A.*, **72**, 4876.
Keller, W., Muller, U., Eicken, I., Wendel, I. and Zentgraf, H. (1978). *Cold Spring Harb. Symp. quant. Biol.*, **42**, 227.
Kelley, R. I. (1973). *Biochem. biophys. Res. Commun.*, **54**, 1588.
Klevan, L. and Crothers, D. M. (1977). *Nucl. Acids Res.*, **4**, 4077.
Kornberg, R. D. (1974). *Science*, **184**, 868.
Kornberg, R. D. (1977). *A. Rev. Biochem.*, **46**, 931.
Kornberg, R. D. and Thomas, J. O. (1974). *Science*, **184**, 865.
Kuo, M. T., Sahasrabuddhe, C. G. and Saunders, G. F. (1976). *Proc. natn. Acad. Sci. U.S.A.*, **73**, 1572.
Lacy, E. and Axel, R. (1975). *Proc. natn. Acad. Sci.U.S.A.*, **72**, 3978.
Lakowicz, J. R. and Weber, G. (1973). *Biochemistry*, **12**, 4171.
Lamb, M. M. and Daneholt, B. (1979). *Cell*, **17**, 835.
Langmore, J. P. and Wooley, J. C. (1975). *Proc. natn. Acad. Sci. U.S.A.*, **72**, 2691.
Lau, A. F. and Ruddon, R. W. (1977). *Exp. Cell Res.*, **107**, 35.
Leffak, I. M., Grainger, R. and Weintraub, H. (1977). *Cell*, **12**, 837.
Levine, Y. K., Birdsall, N. J. M., Lee, A. G. and Metcalf, J. C. (1972). *Biochemistry*, **11**, 1416.
Levitt, M. (1978). *Proc. natn. Acad. Sci. U.S.A.*, **75**, 640.
Levy, A. and Jakob, K. M. (1978). *Cell*, **14**, 259.
Levy-W., B. and Dixon, G. H. (1978*a*). *Nucl. Acids Res.*, **5**, 4155.
Levy-W., B. and Dixon, G. H. (1978*b*). *Can. J. Biochem.*, **56**, 480.
Lilley, D. M. J. (1977). In *Molecular Spectroscopy* (ed. A. R. West), Heyden, London, p. 96.
Lilley, D. M. J., Howarth, O. W., Clark, V. M., Pardon, J. F. and Richards, B. M. (1975). *Biochemistry*, **14**, 4590.
Lilley, D. M. J., Howarth, O. W., Clark, V. M., Pardon, J. F. and Richards, B. M. (1976). *FEBS Lett.*, **62**, 7.
Lilley, D. M. J., Jacobs, M. F. and Houghton, M. (1979). *Nucl. Acids Res.*, **7**, 377.
Lilley, D. M. J. and Pardon, J. F. (1979). *A. Rev. Genet.*, **13**, 197.
Lilley, D. M. J., Pardon, J. F. and Richards, B. M. (1977). *Biochemistry*, **16**, 2853.
Lilley, D. M. J., Richards, B. M., Pardon, J. F., Cotter, R. I. and Worcester, D. L. (1978). In *Gene Expression, 11th FEBS Meeting, Copenhagen* (ed. B. F. C. Clark), Pergamon Press, Oxford, New York, p. 213.
Lilley, D. M. J. and Tatchell, K. (1977). *Nucl. Acids Res.*, **4**, 2039.
Liu, L. F. and Wang, J. C. (1978). *Cell*, **15**, 979.
Lohr, D., Corden, J., Tatchell, K., Kovacic, R. T. and Van Holde, K. E. (1977). *Proc. natn. Acad. Sci. U.S.A.*, **74**, 79.

Lohr, D., Kovacic, R. T. and Van Holde, K. E. (1977). *Biochemistry*, **16**, 463.

Lohr, D., Tatchell, K. and Van Holde, K. E. (1977). *Cell*, **12**, 829.

Lutter, L. C. (1977). *J. molec. Biol.*, **117**, 53.

Lutter, L. C. (1978). *J. molec. Biol.*, **124**, 391.

Lutter, L. C. (1979). *Nucl. Acids Res.*, **6**, 41.

Luzzati, V. and Nicolaieff, A. (1963). *J. molec. Biol.*, **7**, 142.

McKnight, S. L. and Millar, O. L. (1977). *Cell*, **12**, 795.

Martinson, H. G., True, R. J. and Burch, J. B. E. (1979). *Biochemistry*, **18**, 1082.

Mayfield, J. E., Serunian, L. A., Silver, L. M. and Elgin, S. C. R. (1978). *Cell*, **14**, 539.

Millar, O. L. and Beatty, B. R. (1969). *Science*, **164**, 955.

Mirzabekov, A. D., Sanko, D. F., Kolchinsky, A. M. and Melnikova, A. F. (1977). *Eur. J. Biochem.*, **75**, 379.

Mirzabekov, A. D., Shick, V. V., Belyavsky, A. V. and Bavykin, S. G. (1978). *Proc. natn. Acad. Sci. U.S.A.*, **75**, 4184.

Morris, N. R. (1976). *Cell*, **8**, 357.

Moss, T., Cary, P. D., Abercrombie, B. D., Crane-Robinson, C. and Bradbury, E. M. (1976a). *Eur. J. Biochem.*, **71**, 337.

Moss, T., Cary, P. D., Crane-Robinson, C. and Bradbury, E. M. (1976b). *Biochemistry*, **15**, 2261.

Moss, T., Stevens, R. M., Crane-Robinson, C. and Bradbury, E. M. (1977). *Nucl. Acids Res.*, **4**, 2477.

Murphy, R. F., Wallace, R. B. and Bonner, J. (1978). *Proc. natn. Acad. Sci. U.S.A.*, **75**, 5903.

Newrock, K. M., Alfageme, C. R., Nardi, R. V. and Cohen, L. H. (1978). *Cold Spring Harb. Symp. quant. Biol.*, **42**, 421.

Noll, M. (1974a). *Nature, Lond.*, **251**, 249.

Noll, M. (1974b). *Nucl. Acids Res.*, **1**, 1573.

Noll, M. (1977). *J. molec. Biol.*, **116**, 49.

Noll, M. and Kornberg, R. D. (1977). *J. molec. Biol.*, **109**, 393.

Noll, M., Thomas, J. O. and Kornberg, R. D. (1975). *Science*, **187**, 1203.

Olins, A. L. (1978). *Cold Spring Harb. Symp. quant. Biol.*, **42**, 325.

Olins, A. L., Breillatt, J. P., Carlson, R. D., Senior, M. B., Wright, E. B. and Olins, D. E. (1977a). In *The Molecular Biology of the Mammalian Genetic Apparatus*, (ed. P. O. P. Ts'o), Elsevier/North Holland, Amsterdam, p. 211.

Olins, D. E., Bryan, P. N., Harrington, R. E., Hill, W. E. and Olins, A. L. (1977b). *Nucl. Acids Res.*, **4**, 1911.

Olins, A. L. and Olins, D. E. (1974). *Science*, **183**, 330.

Oudet, P., Germond, J. E., Bellard, M., Spadafora, C. and Chambon, P. (1978). *Phil. Trans. R. Soc. B.*, **283**, 241.

Oudet, P., Gross-Bellard, M. and Chambon, P. (1975). *Cell*, **4**, 281.

Palmiter, R. D., Mulvihill, E. R., McKnight, G. S. and Senear, A. W. (1978). *Cold Spring Harb. Symp. quant. Biol.*, **42**, 639.

Pardon, J. F., Cotter, R. I., Lilley, D. M. J., Worcester, D. L., Campbell, A. M., Wooley, J. C. and Richards, B. M. (1978). *Cold Spring Harb. Symp. quant. Biol.*, **42**, 11.

Pardon, J. F. and Richards, B. M. (1979). In *The Cell Nucleus* (ed. H. Busch), vol. 7, p. 37.

Pardon, J. F., Richards, B. M. and Cotter, R. I. (1974). *Cold Spring Harb. Symp. quant. Biol.*, **38**, 75.

Pardon, J. F. and Wilkins, M. H. F. (1972). *J. molec. Biol.*, **68**, 115.

Pardon, J. F., Worcester, D. L., Wooley, J. C., Cotter, R. I., Lilley, D. M. J. and Richards, B. M. (1977). *Nucl. Acids Res.*, **4**, 3199.

Pardon, J. F., Worcester, D. L., Wooley, J. C., Tatchell, K., Van Holde, K. E. and Richards, B. M. (1975). *Nucl. Acids Res.*, **2**, 2163.

Paulson, J. R. and Laemmli, U. K. (1977). *Cell*, **12**, 817.
Pfeiffer, W., Hörz, W., Igó-Kemenes, T. and Zachau, H. G. (1975). *Nature, Lond.*, **258**, 450.
Pinon, R. and Salts, Y. (1977). *Proc. natn. Acad. Sci. U.S.A.*, **74**, 2850.
Pooley, A. S., Pardon, J. F. and Richards, B. M. (1974). *J. molec. Biol.*, **85**, 533.
Porod, G. (1951). *Kolloid Z.*, **2**, 83.
Prunell, A. and Kornberg, R. D. (1978). *Phil. Trans. R. Soc. B.*, **283**, 269.
Renz, M., Nehls, P. and Hozier, J. (1977). *Proc. natn. Acad. Sci. U.S.A.*, **74**, 1879.
Riggs, M. G., Whittaker, R. G., Neumann, J. R. and Ingram, V. M. (1977). *Nature, Lond.*, **268**, 462.
Richards, B. M., Cotter, R. I., Lilley, D. M. J., Pardon, J. F., Wooley, J. C. and Worcester, D. L. (1976). *Current Chromosome Research* (eds. K. Jones and P. E. Brandham), North Holland, Amsterdam.
Richards, B. M. and Pardon, J. F. (1970). *Exp. Cell Res.*, **62**, 184.
Richards, B. M., Pardon, J. F., Lilley, D. M. J., Cotter, R. I., Wooley, J. C. and Worcester, D. L. (1977). *Cell Biol. Int. Rep.*, **1**, 107.
Riley, D. and Weintraub, H. (1978). *Cell*, **13**, 281.
Ris, H. and Kubai, D. F. (1970). *A. Rev. Genet.*, **4**, 263.
Roark, D. E., Geoghegan, T. E. and Keller, G. H. (1974). *Biochem. biophys. Res. Commun.*, **59**, 542.
Sasisekharan, V., Pattabiraman, N. and Gupta, G. (1978). *Proc. natn. Acad. Sci. U.S.A.*, **75**, 4092.
Seale, R. L. (1975). *Nature, Lond.*, **255**, 247.
Seale, R. L. (1978). *Proc. natn. Acad. Sci. U.S.A.*, **75**, 2717.
Sealy, L. and Chalkley, R. (1978). *Nucleic Acids Res.*, **5**, 1863.
Sedat, J. and Manuelidis, L. (1978). *Cold Spring Harb. Symp. quant. Biol.*, **42**, 331.
Seligy, V. L. and Poon, N. H. (1978). *Nucl. Acids Res.*, **5**, 2233.
Shaw, B. R., Herman, T. M., Kovacic, R. T., Beaudreau, G. S. and Van Holde, K. E. (1976). *Proc. natn. Acad. Sci. U.S.A.*, **73**, 505.
Shaw, B. R. and Schmitz, K. S. (1976). *Biochem. biophys. Res. Commun.*, **73**, 224.
Simon, R. H., Camerini-Otero, R. D. and Felsenfeld, G. (1978). *Nucl. Acids Res.*, **5**, 4805.
Simpson, R. T. (1978). *Cell*, **13**, 691.
Simpson, R. T. and Whitlock, J. P. (1976). *Cell*, **9**, 347.
Sobell, H. M., Tsai, C., Gilbert, S. G., Jain, S. C. and Sakore, T. D. (1976). *Proc. natn. Acad. Sci. U.S.A.*, **73**, 3068.
Sollner-Webb, B. and Felsenfeld, G. (1977). *Cell*, **10**, 537.
Sollner-Webb, B., Melchior, W. and Felsenfeld, G. (1978). *Cell*, **14**, 611.
Spadafora, C., Bellard, M., Compton, J. L. and Chambon, P. (1976). *FEBS Lett.*, **69**, 281.
Sperling, L. and Tardieu, A. (1976). *FEBS Lett.*, **64**, 89.
Spiker, S. and Isenberg, I. (1978). *Cold Spring Harb. Symp. quant. Biol.*, **42**, 157.
Staynov, D. Z., Spencer, M., Allan, J. and Gould, H. J. (1979). *Nature, Lond.*, **279**, 263.
Stein, A., Bina-Stein, M. and Simpson, R. T. (1977). *Proc. natn. Acad. Sci. U.S.A.*, **74**, 2780.
Steinmetz, M., Streeck, R. E. and Zachau, H. G. (1975). *Nature, Lond.*, **258**, 447.
Stockley, P. G. and Thomas, J. O. (1979). *FEBS Lett.*, **99**, 129.
Stratling, W. H., Muller, U. and Zentgraf, H. (1978). *Exp. Cell Res.*, **117**, 301.
Suau, P., Kneale, G. G., Braddock, G. W., Baldwin, J. P. and Bradbury, E. M. (1977). *Nucl. Acids Res.*, **4**, 3769.
Sussman, J. L. and Trifonov, E. N. (1978). *Proc. natn. Acad. Sci. U.S.A.*, **75**, 103.
Tata, J. R. and Baker, B. (1978). *J. molec. Biol.*, **118**, 249.
Tatchell, K. and Van Holde, K. E. (1978). *Proc. natn. Acad. Sci. U.S.A.*, **75**, 3583.
Thoma, F. and Koller, T. (1977). *Cell*, **12**, 101.

Thomas, G. J., Prescott, B. and Olins, D. E. (1977). *Science,* **197**, 385.

Thomas, J. O. and Butler, P. J. G. (1977). *J. molec. Biol.,* **116**, 769.

Thomas, J. O. and Furber, V. (1976). *FEBS Lett.,* **66**, 274.

Thomas, J. O. and Kornberg, R. D. (1975). *Proc. natn. Acad. Sci. U.S.A.,* **72**, 2626.

Thomas, J. O. and Thompson, R. J. (1977). *Cell,* **10**, 633.

Todd, R. D. and Garrard, W. T. (1977). *J. Biol. Chem.,* **252**, 4729.

Trifonov, E. N. and Bettecken, T. (1979). *Biochemistry,* **18**, 454.

Tsanev, R. and Petrov, P. (1976). *J. Microsc. Biol. Cell,* **27**, 11.

Van Der Westhuyzen, D. R. and Von Holt, C. (1971). *FEBS Lett.,* **14**, 333.

Van Holde, K. E., Sahasrabuddhe, C. G. and Shaw, B. R. (1974). *Nucl. Acids Res.,* **1**, 1579.

Van Holde, K. E. and Weischet, W. O. (1978). In *The Cell Nucleus* (ed. H. Busch), vol. IV, Academic Press, New York, p. 75.

Varshavsky, A. J. and Bakayev, V. V. (1975). *Mol. Biol. Rep.,* **2**, 247.

Varshavsky, A. J., Bakayev, V. V. and Georgiev, G. P. (1976). *Nucl. Acids Res.,* **3**, 477.

Vidali, G., Boffa, L. C. and Allfrey, V. G. (1977). *Cell,* **12**, 409.

Vidali, G., Boffa, L. C., Bradbury, E. M. and Allfrey, V. G. (1978). *Proc. natn. Acad. Sci. U.S.A.,* **75**, 2239.

Vinograd, J. and Lebowitz, J. (1966). *J. gen. Physiol.,* **49**, 103.

Von Holt, C., Strickland, W. N., Brandt, W. F. and Strickland, M. S. (1979). *FEBS Lett.,* **100**, 201.

Wang, J. C. (1978). *Proc. natn. Acad. Sci. U.S.A.,* **76**, 200.

Weintraub, H. and Groudine, M. (1976). *Science,* **193**, 848.

Weintraub, H., Palter, K. and Van Lente, F. (1975). *Cell,* **6**, 85.

Weintraub, H. and Van Lente, F. (1974). *Proc. natn. Acad. Sci. U.S.A.,* **71**, 4249.

Weintraub, H., Worcel, A. and Alberts, B. (1976). *Cell,* **9**, 409.

Weisbrod, S. and Weintraub, H. (1979). *Proc. natn. Acad. Sci. U.S.A.,* **76**, 630.

Weischet, W. O., Tatchell, K., Van Holde, K. E. and Klump, H. (1978). *Nucl. Acids Res.,* **5**, 139.

Whitlock, J. P., Rushizky, G. W. and Simpson, R. T. (1977). *J. Biol. Chem.,* **252**, 3003.

Whitlock, J. P. and Simpson, R. T. (1977). *J. Biol. Chem.,* **252**, 6516.

Whitlock, J. P. and Stein, A. (1978). *J. Biol. Chem.,* **253**, 3857.

Wilkins, M. H. F., Zubay, G. and Wilson, H. R. (1959). *J. molec. Biol.,* **1**, 179.

Woodcock, C. L. F. (1973). *J. Cell. Biol.,* **59**, 368a.

Wooley, J. C. and Langmore, J. P. (1978). In *Molecular Human Cytogenetics* (eds. R. S. Sparkes, D. Comings and C. F. Fox), Academic Press, New York, p. 41.

Worcel, A. and Burgi, I. E. (1972). *J. molec. Biol.,* **71**, 127.

Worcel, A., Han, S. and Wong, M. L. (1978). *Cell,* **15**, 969.

Wray, W., Mace, M., Daskal, Y. and Stubblefield, E. (1978). *Cold Spring Harb. Symp. quant. Biol.,* **42**, 361.

Zama, M., Bryan, P. N., Harrington, R. E., Olins, A. L. and Olins, D. E. (1978*a*). *Cold Spring Harb. Symp. quant. Biol.,* **42**, 31.

Zama, M., Olins, D. E., Prescott, B. and Thomas, G. J. (1978*b*). *Nucl. Acids Res.,* **5**, 3881.

8

Oligonucleotide and polynucleotide – drug complexes in the crystalline state

Stephen Neidle

INTRODUCTION

It has been known for many years that various molecules possessing a planar aromatic grouping as a common structural feature can interact with genetic material. Thus, acridine orange, in particular, has long been used as a vital stain in cytological procedures, as have various other acridine derivatives with antibacterial properties. It is now well established that numerous such molecules, many of which are of medicinal interest and use, exert their major *in vivo* effects by direct DNA inter-action, thereby interfering with processes of nucleic acid expression and synthesis (see, for example, Gale *et al.*, 1972). Interaction with ribonucleic acids, both single and double stranded has also been implicated in the action of these druges (see, for example, Finkelstein and Weinstein, 1967).

The advent of the double-helical model for DNA structure provided the major impetus for physicochemical and structural studies on these drug–DNA complexes. These culminated in the intercalation model of Lerman (1961) (figure 8.1), which satisfactorily explains many of the gross structural changes observed on DNA binding (Gale *et al.*, 1972; Neidle, 1979). The central feature of this hypothesis is the insertion of a planar drug chromophore in between two adjacent base pairs, in order to increase their separation from 3.4 to 6.8 Å. It is a feature of the model, and of most physical investigations of intercalation until recently, that attention has almost entirely focussed on the alterations in base-pair disposition. This is unsur-prising in view of the inability of most methodology to probe the fine structural changes which must occur in the phosphodiester-sugar backbone concomitant with the separation of base pairs. In biological terms these former changes are of some importance, bearing in mind that many cellular events relevant to drug and mutagen action, such as DNA repair, doubtless initially involve recognition of backbone geometry and abnormalities in such geometry. It will be shown in this review that although the currently available structural data fully support the essential base-pair-

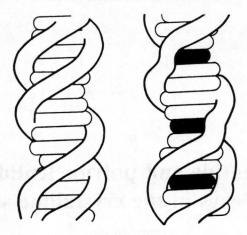

Figure 8.1   The intercalation model of Lerman (1961). The left-hand side represents native DNA with the base pairs as end-on viewed discs. The right-hand side represents drug (shaded discs) intercalated in between base pairs.

separating feature of the classical intercalation model, it does not enable the back-bone conformational changes to be unequivocally defined; thus, to date, we are unable to answer most of the major stereochemical questions posed by the inter-calation phenomenon:

(1) What are the changes in nucleic acid geometry necessary to produce inter-calation, both in DNA and RNA?

(2) Are these changes dependant on the nature of the drug?

(3) How do these changes relate to the observations of unwinding, neighbour exclusion and sequence-preference binding?

STRUCTURAL STUDIES ON DRUG–POLYNUCLEOTIDE COMPLEXES

Fibre diffraction analyses

The relatively well-ordered semicrystalline fibres obtainable for DNA and RNA in general produce x-ray diffraction diagrams which are of relatively high resolution and information content. This has enabled their (averaged) structures to be well defined by computerised model-fitting procedures (see chapter 4). However, fibres of drug-polynucleotide complexes do not show such a high degree of order, and thus their diffraction patterns have few maxima. This inevitably means that the level of reliable structural data obtainable from them is strictly limited.

Diffraction patterns from fibres of drug–DNA complexes generally show evidence of little or any long-range order. However, Lerman (1961) showed that the crucial 3.4 Å base stacking reflection was still present. This provided strong evidence for the intercalation model, which retains 3.4 Å parallel spacing between chromophore and base pairs. Examination of molecular models suggested that the base pairs either side of the intercalated acridine were mutually oriented at $-9°$; they had been unwound by $45°$ in a left-handed direction from the B-DNA value of $36°$. Subsequent physical and chemical studies (Lerman, 1963; 1964a, b) confirmed the essential correctness of the intercalation concept, although the unwinding angle was modified to $36°$. A low-angle x-ray scattering study (Luzzati, Masson and Lerman, 1961) also provided results for the DNA-proflavine complex consistent with the intercalation model, with the complex having a diminished mass per unit length compared to the nucleic acid alone.

A detailed study of DNA complexes with acridine orange and proflavine (Neville and Davies, 1966) produced diffraction patterns which, although showing layer and line disorder, still had some interpretable detail. These authors noted that satisfactory molecular models for intercalation may be produced with unwinding angles between $12°$ and $45°$ per intercalated drug, although their fitting of optical transforms to diffraction patterns were performed assuming an unwinding of $36°$. Interpretation of diffraction patterns from DNA-ethidium fibres using molecular model-building techniques (Fuller and Waring, 1964) suggested that only an unwinding of $12°$ per drug intercalation was necessary. The relatively detailed molecular model produced in this study showed the ethidium molecule inserted via the major groove and stabilised by hydrogen bonds between its amino groups and the backbone phosphate oxygen atoms. This value for the unwinding angle has been widely adopted (Gale et al., 1972) as a standard; the unwinding induced in a closed circular DNA by other intercalating drugs has been often calculated relative to the $12°$ figure. Molecular models have been proposed from the interpretation of diffraction patterns of complexes between DNA and the anticancer drug daunomycin (Pigram, Fuller and Hamilton, 1972), and actinomycin (Hamilton, Fuller and Reich, 1963). In the latter case, the proposals involve nonintercalative binding.

More recently, fibres of DNA complexed with a planar, platinum-containing, intercalative compound, have been examined (Bond et al., 1975). These produced diffraction patterns interpretable in terms of 10.2 Å repeating units. Since platinum atoms dominate the total electron density, it has been concluded that the data strongly supports a 'neighbour exclusion' structural model, with the intercalator binding maximally at every other potential site (Cairns, 1962). Bond et al. conclude from model-building studies that their excluded-site model has an unwinding angle of $22 \pm 6°$ per intercalated dinucleotide. They were unable with their diffraction data, to distinguish between three possible detailed conformation models for inter-calation, although the one with alternating C2$'$ $endo$ − C3$'$ $endo$ sugar pucker (at the binding site) did not have good agreement between observed and calculated Fourier transforms. Indeed none of the models satisfactorily explained the observed patterns, possibly on account of disordering effects.

The singular lack of progress in establishing intercalative polynucleotide geometry from fibre patterns is in marked contrast with the situation for the native polymers (see chapter 4). It is likely that many effects such as disorderings and helix distortions contribute to the lack of precise repetition in the natural nucleic acid complexes, and hence to their poor quality diffraction patterns. These problems may be overcome in the future by the use of synthetic, defined-sequence polymers.

Theoretical models of intercalation into nucleic acids

The techniques of linked-atom, best-fit, molecular model-building (Smith and Arnott, 1978), have been used to calculate plausible models for proflavine intercalation into both *A* and *B* forms of DNA. In both cases, the model system used was a duplex tetranucleotide. Thus, the residues adjacent to the intercalation site were conformationally defined as well as those primarily involved at the site. All residues were constrained to be cohelical.

In the case of *B*-DNA (Alden and Arnott, 1975), two models were produced; one with alternating C3' *endo* -(3',5')-C3' *exo* sugar pucker at the intercalation site being sterically more favourable than the other with all C3' *exo* (figure 8.2). The adjacent residues do not change markedly from *B*-DNA conformations; the major changes are in the sugar pucker and in torsion angles γ and ϵ at the intercalation site (table 8.1). Both adopt *trans* values. The total turn angles of 90°, distributed over the three residues, correspond to an unwinding at the intercalation site of about −18°; however, Alden and Arnott consider this total angular change over the affected region to be the more meaningful value. This model has the amino groups of the proflavine molecule pointing towards the major groove of the DNA. The nucleotide conformation is such as to exclude hydrogen-bonding stabilisation of

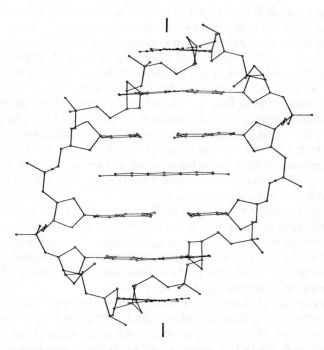

Figure 8.2 The optimum computed stereochemistry for the *B*-DNA–proflavine complex, viewed from the minor groove. The vertical lines represent the axis of the polynucleotide double helix. (From Alden and Arnott, 1975.)

Table 8.1 Conformational angles for intercalated DNA sites from theoretical
model building, with constrained helix linearity

	$\chi(5')$	α	β	γ	δ	ϵ	$\chi(3')$
For A-DNA[a]	23	211	261	201	212	149	23
For B-DNA[b]	52	177	273	192	180	187	52

[a]Alden and Arnott (1977).

[b]Alden and Arnott (1975).

Table 8.3 provides data for the native polynucleotides.

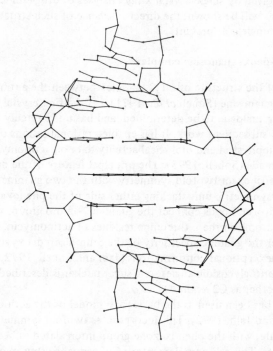

Figure 8.3 The computed optimum stereochemistry for the A-DNA–proflavine complex,
viewed facing the major groove. (From Alden and Arnott, 1978.)

the drug's exocyclic amino groups with the charged phosphate oxygen atoms of
the backbone.

A very similar approach has been adopted to the analysis of models for an A-DNA
proflavine complex (figure 8.3) (Alden and Arnott, 1977). A unique unstrained
solution was obtained, with torsion angles γ and ϵ again in *trans* conformations, but
(in contrast to the B-DNA complex) with all sugars retaining their native A-DNA C3′
endo pucker. Furthermore, the total turn angle over all three residues was found to

be 96.6°, corresponding to an unwinding at the central site of only −1.6°. Again in contrast to their *B*-DNA-proflavine model, Alden and Arnott find that the drug complex can be further stabilised by exocyclic hydrogen bonds to both backbone strands simultaneously. It is considered that a proflavine–RNA complex will share many features with this structure; some aspects of the proflavine-ribodinucleoside crystal structure (Neidle *et al.*, 1977) certainly bear out this prediction.

SINGLE CRYSTAL STUDIES ON DRUG–OLIGONUCLEOTIDE COMPLEXES

We have seen in the preceeding sections that analyses of drug–polynucleotide complexes have not yet provided atomic-level data. To some extent, such information has, and is, being given by single-crystal x-ray analyses of drug–dinucleoside complexes; however, as will be shown, the direct relevance of such structures to nucleic acid complexes is unclear at present.

The actinomycin–deoxyguanosine complex

The elucidation of the structure of a 1:2 complex between the antibiotic actinomycin and deoxyguanosine (Sobell *et al.*, 1971) was the first crystal structure of any drug–receptor analogue to be determined, and has undoubtedly provided the impetus for much subsequent work. It has enabled rationalisations of a large body of biochemical, biophysical and biological activity data on actinomycin and its derivatives to be made (Sobell, 1973). The principal feature of the nucleoside complex is its almost exact twofold symmetry with the two guanine bases in the crystallographic asymmetric unit stacking either side of the phenoxazone chromophors. Strong hydrogen bonds connect the guanine 2-amino groups with the carbonyl oxygen atoms on the L-threonine residues of actinomycin, thus providing an explanation for the guanine specificity of the drug. Both deoxyguanosine residues adopt unexceptional conformations (Jain and Sobell, 1972; Sobell and Jain, 1972) with *anti* glycosidic angles; one sugar pucker is described as C3′ *endo*, C2′ *exo*, and the other as C2′ *endo*, C3′ *exo*.

This structure has been used as the basis for a model of the actinomycin–DNA complex (Sobell and Jain, 1972). This incorporates twofold symmetry at the intercalative binding site, with the phenoxazone group intercalated in between two dG − dC base pairs. This self-complementary G.C sequence thus maximises the guanine-threonine hydrogen-bonding requirement. The model was constructed with the hexanucleotide sequence d(ApTpGpCpApT), and has the central base pairs unwound by 18° and the terminal A.T ones by 8°. Thus the total unwinding angle produced by actinomycin is (18 + 2 × 8), that is, 34°. The sugar puckers at the intercalation site are alternating C3′ *endo*-(3′,5′)-C2′ *endo*. Several of the backbone torsion angles in this model assume unexpected near eclipsed values, for example the ε angle for the deoxyguanosine residue is only 8.8°. As pointed out by Alden and Arnott (1975), such factors together with distortions from normality noticed in several bond and torsion angles, suggest that the model as a whole has considerable steric strain. A revised version of this model has recently been made (Sobell *et al.*, 1977). The modifications involving departures from exact diad symmetry together with other backbone distortions, go some way to answering these objections.

Drug–dinucleoside crystal structures

The demonstration (Seeman *et al.*, 1976, Rosenberg *et al.*, 1976) that the ribo-dinucleoside phosphates GpC and ApU form self-complementary duplexes in their crystal structures (with many features of RNA being displayed) led directly to the hypothesis that drug–dinucleoside structures would be relevant to drug–polynucleotide complexes. This notion has been pursued in a number of laboratories, so that some twelve such structures have been determined to date.

The ApU -9- aminoacridine complex (Seeman, Day and Rich, 1975) is not of the base-paired duplex type, but has Hoogsteen base pairs (Voet and Rich, 1970) connecting ApU units to form infinite arrays in the crystal. This further illustrates the concept (detailed in chapter 7) that intercalating drugs preferentially bind to pyrimidine-(3', 5')-purine sequences, and that other sequences become duplex-destabilised, at least when they are complexed in short fragments of oligomer.

Table 8.2 gives some pertinent information for a number of Watson-Crick base-paired duplex (double-helical-fragment) structures. It is important to note that many of these unfortunately do not give high-resolution diffraction data. Thus, the fine details of molecular geometry reported for such structures need to be analysed with considerable caution.

Furthermore, some of these structures are very sensitive to difference in crystallographic refinement procedures; in the case of deoxy (CpG)-proflavine the mutual twist of bases in the base pairs altered by $10°$ between intermediate and final refinements (Neidle and Berman, unpublished observations). Therefore, it would appear wise to guard against possible over-interpretation of structural parameters for, especially, the lower resolution structures. Conclusions based on only minor deviations from normality need particularly careful scrutiny.

Ethidium has been found to form crystalline complexes with both 5-iodo-UpA (Tsai, Jain and Sobell, 1975*a, b*; 1977) and 5-iodo-CpG (Jain, Tsai and Sobell, 1977) which are closely similar. These were the first structures to conclusively demonstrate at atomic resolution drug intercalation in between adjacent Watson-Crick base pairs. Figure 8.4 shows two views of the 5-iodo-CpG, one with base pairs 6.8 Å apart, and the complex as a whole having pseudo twofold symmetry. Both structures have alternating C3' *endo*-(3', 5')-C2' *endo* sugar puckers on individual dinucleoside strands, with the ethidium molecule intercalated from the minor groove, and a resultant base-turn angle* between upper and lower base pairs of about $10°$.

---

*Defined as the angle between the vectors connecting C1' atoms of each base pair when projected on the average base plane viewed from a point perpendicular to this plane.

Table 8.2 Crystallographic data on some drug-dinucleoside complexes

Complex	Space group	Contents of crystallographic asymmetric unit	Resolution limit of data (Å)	Crystallographic residual	Average standard deviation of bond lenths (Å)	Reference
Iodo CpG-ethidium	$P2_1$	2 ethidium, i^{5}CpG, 27H$_2$O, 4 Methanol	1.14	0.158	0.06	Jain, Tsai and Sobell (1977)
Iodo UpA-ethidium	C2	2 ethidium, i^{5}UpA, 27H$_2$O	1.34	0.20	0.1	Tsai, Jain and Sobell (1977)
Iodo CpG-9 aminoacridine	$P2_1$	4 9-aminoacridine, i^{5}CpG, 21H$_2$O	1.31	0.147	0.1	Sakore et al. (1977) Sakore, Reddy and Sobell (1979)
Iodo CpG-proflavine	C2	2 proflavine, i^{5}CpG, 15H$_2$O, 1 Methanol	1.31	0.202	0.1	Reddy et al. (1979)
CpG-proflavine	C2	1½ proflavine, 1 CpG, 12H$_2$O, ½SO$_4^{2-}$	0.85	0.102	0.02	Neidle et al. (1977) Berman et al. (1979)
CpG-acridine orange	P1	1 acridine orange, 2 CpG, 1 Na$^+$, 24H$_2$O	0.85	0.095	?	Wang, Quigley and Rich (1979)
d(CpG)-platinum terpyridine	$P2_12_12_1$	2 platinum terpyridine, 2d(CpG), 2OH$_2$O	1.1	0.15	?	Wang et al. (1978)
d(CpG)-proflavine	$P2_12_12$	2 proflavine, 2d(CpG), 27H$_2$O	0.89	0.15	0.04	Shieh et al. (1980)
ApA-proflavine	$P2_12_12$	2 proflavine, 1 ApA, 15H$_2$O$_2$ 1 SO$_4^{2-}$	0.89	0.106	0.03	Neidle et al. (1978)
CpA-proflavine	$P4_22_12$	1 proflavine, 1 CpA, ? H$_2$O	0.8	0.117	0.02	Westof and Sundaralingam (1980)

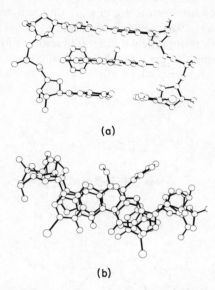

(a)

(b)

Figure 8.4   Two views of the ethidium: 5-iodo-CpG complex (redrawn from Jain, Tsai and Sobell, 1977): (a) looking along the mean planes of the base pairs, (b) looking down on to these planes.

This structural data has been evaluated in terms of intercalation into DNA itself (Sobell *et al.*, 1977). The 10° base-turn angle in the ethidium-dinucleoside structures is said to correspond to a −26° unwinding in the polymer, which agrees remarkably well with an estimate from alkaline titration studies on supercoiled closed-circular DNA (Wang, 1974). In both model structures, a small (approximately 8°) angular twist between base pairs has been found, which, it has been concluded, reflects the intercalative process. In DNA and RNA, it is argued, there is bending or 'kinking' in order for an intercalator to bind, these kinks being produced by alternation in sugar pucker to mixed C3′ *endo*-(3′,5′)-C2′ *endo*. It was also observed in the dinucleoside complexes that a 'helical screw axis dislocation' was present, with the result that helical axes in intercalated DNA are no longer colinear on either side of the inter-calation site. Such concepts have enabled (Sobell *et al.*, 1977) molecular models of DNA-ethidium, -actinomycin and -ihrediamine to be constructed. Further ramifications of the kinking concept (Sobell, Lozansky and Lessen, 1978), in relation to chromatin structure and DNA breathing and dynamics are outside the scope of this review.

The structure of the 9-aminoacridine 5-iodo-CpG complex (Sakore *et al.*, 1977; Sakore, Reddy and Sobell, 1979), contains two distinct intercalated CpG duplexes. In one the intercalation is symmetric, whilst in the other it is markedly asymmetric. In both cases the base-turn angles are small (8–10°). Complexes of 5-iodo-CpG with ellipticine, 3.5.6.8-tetramethyl-N-methyl phenanthrolinium (Jain, Bhandary and Sobell, 1979) and acridine orange (Reddy *et al.*, 1979) all show small base-turn angle and mixed sugar pucker characteristics, as well as having 2:2 drug: dinucleoside ratios. The high-resolution structure of a 2:1 CpG:acridine orange complex (Wang, Quigley and Rich, 1979) again shows mixed sugar pucker, with a 10° base turn angle. This analysis is of particular interest in respect of its well-documented solvent structure, an aspect of the intercalation phenomenon to which perhaps insufficient attention has been paid.

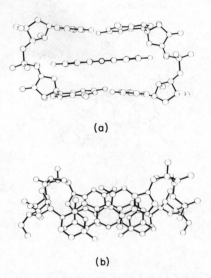

(a)

(b)

Figure 8.5   (a), (b) Two views of the proflavine–CpG complex. The externally stacked proflavines have not been included in view (b), looking on to the planes of the base pairs.

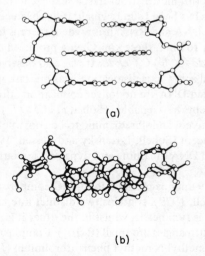

(a)

(b)

Figure 8.6   (a), (b) Two corresponding views of the proflavine–deoxy (CpG) complex.

Aspects of the high-resolution crystal structure of a 3:2 proflavine:CpG complex (Neidle *et al.*, 1977; Berman *et al.*, 1979) are in marked contrast to these features. This structure has a proflavine cation precisely symmetrically intercalated (figure 8.5), with its exocyclic nitrogen amino atoms hydrogen bonded to a phosphate-oxygen atom on each strand, at distances of 3.0 Å. All sugars have C3′ *endo* pucker and the base-turn angle is 34°. These features resemble some of those proposed by Alden and Arnott (1977) for the *A*-DNA–proflavine complex. Although the mutual orientation of the base pairs is thus as in both RNA-11 and uncomplexed dinucleosides, they have a translational shift relative to these nonintercalated situations. The pro-flavine:CpG crystal structure also displays drug binding to the exterior of the phosphodiester backbone, with hydrogen bonding to both phosphate oxygen and cytosine ribose 2′ hydroxyl atoms. The less well-defined crystal structure (table 8.2) of a 2:2 proflavine:5-iodo-CpG complex (Reddy *et al.*, 1979) does not appear to show this external binding, although the intercalation geometry itself is similar to that in the 3:1 one, with near-symmetrical intercalation of drug, all C3′ *endo* sugar puckers and a 36° base-turn angle.

Two drug deoxyribodinucleoside crystal structures have been reported (table 8.2). That between deoxy (CpG) and the intercalating platinum compound 2-hydroxy-ethenethiolato -2,2′,2″-terpyridine-platinum II (Wang *et al.*, 1978) is a 2:2 self-complementary complex with alternating C3′ *endo* -(3′,5′)-C2′ *endo* sugar puckers and a 13° base turn angle. The proflavine-deoxy(CpG) 2:2 complex (Shieh *et al.*, 1980) has some features distinct from the corresponding ribo-structure. Thus the intercalated proflavine is markedly assymmetric with respect to the two base pairs (figure 8.6), and it does not hydrogen bond to phosphate-oxygen atoms of the back-bones. Interestingly, the two strands themselves are asymmetric – one has the commonly observed C3′ *endo* -(3′,5′)-C2′ *endo* mixed sugar pucker pattern, the other is all C3′ *endo*. This structure thus clearly demonstrates the secondary import-ance of sugar pucker in dinucleoside intercalation geometry, a point amplified in the next section.

The crystal structures of two non-self-complementary dinucleoside–drug com-plexes have recently been determined. Both in their different ways, shed new light

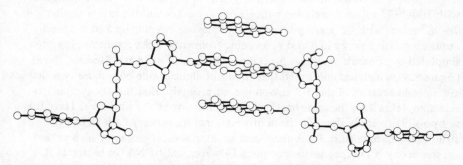

Figure 8.7   The proflavine–ApA complex, showing two base-paired ApA units flanked by two crystallographically nonidentical stacking proflavine cations.

on the conformational behaviour of dinucleosides. That of CpA with proflavine (Westhof and Sundaralingam, 1980) surprisingly forms a symmetrical base-paired dimer with A. . .A (through N6 and N7) and C. . .C (through N3, N4 and O2) base pairs with a disordered proflavine intercalated in between them. Again, the sugar puckers follow the alternating pattern; the conformation of the backbone itself also follows the behaviour of the self-complementary duplex complexes (see below). The 1:2 ApA:proflavine complex (Neidle *et al.*, 1978) adopts an at first sight less surprising extended conformation, with (N6. . .N7) base pairs linking ApA units in the crystal lattice (figure 8.7), and having each base pair sandwiched in between two proflavine cations by 3.4 Å separations. However, a detailed examination shows that the ApA backbone has unusual features; the two halves have very different conformations with the glycosidic linkage at the 3′ end being *syn* (χ of $-120°$) and the C4′-C5′ angle ϵ assuming a *trans* value of 175°. The 5′ ribose adopts the unusual C1′ *exo*, C2′ *endo* conformation; that of the 3′ ribose is the more normal C3′ *endo* pucker. Curiously, the torsion angles round the phosphorus-oxygen bonds (β and γ), usually considered in the 'rigid nucleotide' concept to be the more flexible areas of a nucleotide, adopt almost precisely A-RNA values. It may be surmised that the stacking interactions of the drug are largely responsible for these unexpected conformational effects. Whether or not complexes with single-stranded nucleic acids have similar bizarre behaviour is a matter for further experimentation.

Rationalisations of drug–dinucleoside conformational behaviour and properties

The foregoing descriptions of the various drug–complex crystal structures suggest that for the self-complementary duplex ones there is considerable conformational variability. Table 8.3 shows the principal features of such structures. It is apparent that for all these structures only two angles (δ, about C5′——O5′, and the 3′ glycosidic angle) differ substantially from their values either in A-RNA or in uncomplexed dinucleoside phosphates. Furthermore, the backbone angles in the complexes are confined to narrow ranges of values (figure 8.3). The minor deviations in these torsion angles between various complexes may be ascribed to secondary effects such as differences in crystal packing and stacking forces. (It is not surprising to note that the pattern of $\chi(3′)$ values parallels 3′ end sugar pucker, with 100–117° values correlating with C2′ *endo* pucker and the rather smaller 78–90° values with C3′ *endo* pucker.) The process of opening-up a base-paired duplex by changes only in δ and χ, on going from an A-RNA geometry, has been simulated by computer graphics procedures (Berman, Neidle and Stodola, 1978) (figure 8.8), which has confirmed that for a ribodinucleoside only these two changes are indeed necessary. Table 8.3 also shows, surprisingly, that the deoxydinucleoside complexes have the same backbone conformation as the ribo ones. Thus, it is not possible to easily describe the conformational transition of a B-DNA type structure with base pairs 3.4 Å apart into an intercalation geometry such as that for the deoxy (CpG) proflavine complex. However, an A-DNA (or better an $A′$-DNA) type starting point (table 8.3) allows the δ, χ transformation to be accomplished with ease.

Table 8.3 Conformational angles for dinucleoside phosphate: drug complexes

	α	β	γ	δ	ε	χ(5′)	χ(3′)	Reference
dCpG. Proflavine	210	290	290	219	46	16	80	Shieh et al. (1979)
	203	300	287	218	73	10	113	
dCpG. Platinum complex	201	287	282	226	57	32	114	Wang et al. (1978)
	194	292	308	217	84	34	117	
CpG. Proflavine	204	292	287	234	53	18	87	Neidle et al. (1977)
CpG. Acridine orange	211	301	288	237	50	9	105	Wang, Quigley and Rich (1979)
	225	298	297	226	40	8	115	
i^{5}CpG. Ethidium	226	281	286	210	72	3	101	Jain, Tsai and Sobell (1977)
	225	291	291	224	55	24	109	
i^{5}UpA. Ethidium	207	286	291	236	52	26	99	Tsai, Jain and Sobell (1977)
	218	302	276	230	70	14	100	
Average	211(11)	293(7)	290(8)	226(8)	59(12)	18(11)	103(12)	
Average for dinucleotide duplexes	217(5)	289(4)	290(6)	174(7)	57(7)			Berman and Shieh (1980)
A-RNA	213	281	300	175	49	14	14	Arnott, Smith and Chandrasekaran (1976)
A′-DNA	205	290	299	176	51	13	13	
A-DNA	175	315	270	211	47	27	27	
B-DNA	159	261	321	209	31	85	85	

Two sets of values for a particular complex indicate two dinucleosides in the crystallographic asymmetric unit.

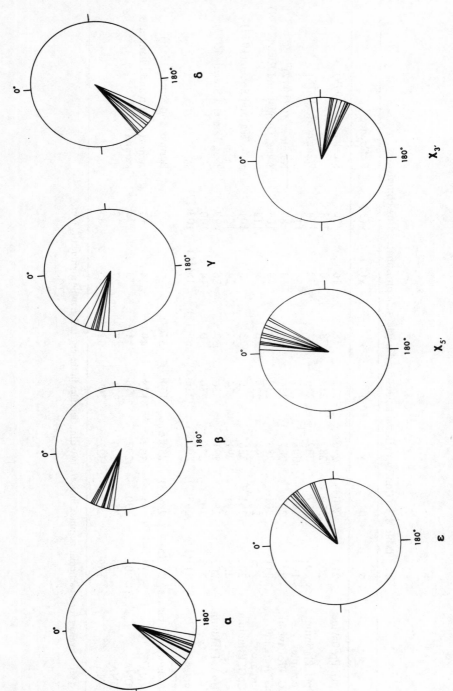

Figure 8.8 The distribution of conformational angles in intercalated dinucleosides.

It has also been shown by computerised model building (Berman, Neidle and Stodola, 1978) that otherwise indistinguishable intercalated dinucleosides, but with differing sugar puckers, can be constructed. Starting with A-RNA geometry, dinucleosides of either all C3$'$ *endo* or C3$'$ *endo*-(3$'$,5$'$)-C2$'$ *endo* pucker were transformed by changing δ and χ. The resultant structures had equivalent backbone angles. Furthermore, it was possible to generate structures with differing base turn angles, yet still with at most minor differences in backbone conformation. It appears that the base-turn angle is not so much dependent on these parameters as on small variations in base-pair geometry, possibly in combination with them. The value adopted for the base turn is probably related to the nature of the intercalating chromophore, with sterically bulky ones such as ethidium necessarily forcing a large angle. It is important to realise that base turn is not equivalent to a classical polymeric unwinding angle, and it may only be fortuitous when this seems to be related to helical unwinding. A not unconnected common misconception is that dinucleosides are 'miniature double helices'; they are merely base-paired dimers which may or may not possess the potentiality of being incorporated into a polynucleotide helix.

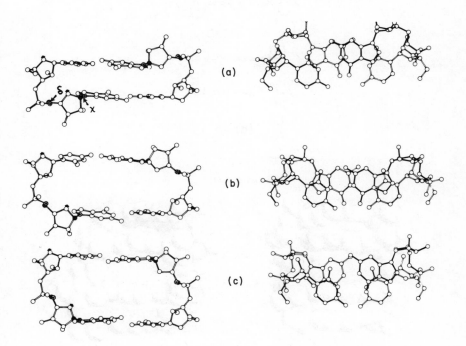

Figure 8.9 The stepwise intercalative stretching of a ribonucleoside phosphate (CpG) (a) in an A-RNA conformation with δ = 175° and χ = 13°, (b) now with δ = 200°, χ = 45°, (c) with δ = 225°, χ = 80°. The arrows represent the vector between the centre of the C2$'$– C3$'$ bond and the O1$'$ atom. (From Berman, Neidle and Stodola, 1978).

Drug–polynucleotide models derived from single-crystal analyses

Mention has already been made of the DNA–ethidium binding models derived from the crystal structures of 5-iodo-CpG and 5-iodo-UpA complexed with this drug (Sobell *et al.*, 1977). Table 8.4 gives the conformational angles for the two models examined by these workers. In both cases only parameters for the intercalated residue plus one nucleotide on either side have been given here, although in the case of the isolated-site model the two residues beyond this tetranucleotide (one on either side) do not have standard *B*-DNA forms. Prominent features of both models are alternating sugar puckers on each strand at the binding sites, accompanied by partial unstacking of base pairs to form kinks, and unwinding of adjacent base pairs. It is noteworthy that neither model incorporates the precise geometry found in the dinucleoside complexes (table 8.2). In particular, γ has been increased by about $40°$, δ decreased by over $30°$ and ϵ decreased by over $15°$. These relaxations of the

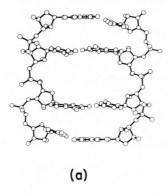

(a)

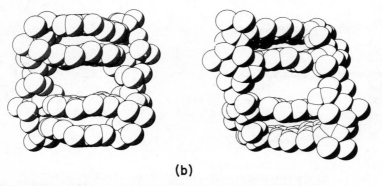

(b)

Figure 8.10   The conformation of the intercalated ribotetranucleotide for the extended exclusion site model (Berman and Neidle, 1979). (a) Ball and stick representation, (b) space-filling representation, with (at right) view rotated by 180°.

constraints imposed by the dinucleoside geometry are the principal reasons for the facile manner in which *B*-DNA residues were added to the (modified) intercalated core site.

The problems of constructing intercalated polynucleotides with the experimentally-determined dinucleoside–drug conformations being constrained have been examined by Berman and Neidle (1979). This study was confined to RNA analogues; only subsequently has the data on deoxydinucleoside complexes become available. Close examination of the intercalated dinucleosides (compare with figure 8.9) reveals that they are asymmetric, as shown by the non-equivalence of 3' and 5' glycosidic angles (and hence by the non-equivalence of the sugar vectors, as shown in figure 8.8). This asymmetry has a profound effect on attempts at polymer generation – a fact recognised earlier (Sobell *et al.*, 1977) in the concept of helical disruption. Thus, because of the high χ angle at the guanosine ends of the CpG duplex, it is not possible to have an *A*-RNA geometry adjacent to the experimentally determined dinucleoside one. It is also not possible to polymerise the geometry so as to have intercalation at every site. Neighbour exclusion is then an inescapable consequence of the χ angle asymmetry, and not of concepts such as mixed sugar pucker. Using both manual and computer model building, a model was constructed

Table 8.4 Conformational angles for the tetranucleotide core of drug-polynucleotide models

I Models of Sobell *et al.* (1977) for DNA-ethidium

(a) The isolated-site (extended-exclusion) model

	$\chi(5')$	α	β	γ	δ	ϵ	$\chi(3')$
Residue 1	69	187	253	278	152	69	29
2	29	223	279	327	193	42	90
3	90	162	237	302	164	72	54

(b) The neighbour-exclusion model

Residue 1	90	180	240	285	152	69	29
2	29	223	279	326	193	42	90
3	90	180	240	285	152	69	29

II Models of Berman and Neidle (1979) for an intercalated ribotetranucleotide.

(a) The isolated-site (extended-exclusion) model

	$\chi(5')$	α	β	γ	δ	ϵ	$\chi(3')$
Residue 1	29	274	210	303	106	64	13
2	13	213	281	300	235	50	85
3	85	178	275	148	210	175	30

(b) The neighbour-exclusion model

Residue 1	85	171	286	143	196	176	13
2	13	213	281	300	235	50	85
3	85	172	288	143	196	176	13

that attempted to revert to normal RNA geometry as near as possible to the binding site. This model (figure 8.10, table 8.4) naturally results in extended-site exclusion, with the next nearest intercalation site being probably four residues distant for a poly (G—C), poly (G—C) sequence with pyrimidine -(3',5')-purine sequence preference binding being observed. The model has a 33° base turn angle for the intercalated dinucleoside itself. However, the actual turn angle of the four residues in the model is much smaller than 3 × 33°. Thus unwinding has only become apparent at the tetranucleotide level, and does not appear to be a function of the base turn angle at the binding site itself.

The second model, that for maximal neighbour exclusion binding (figure 8.11), has symmetrical adjacent site conformations with twisted and bent base pairs. No residue ever assumes a normal RNA conformation (table 8.4), although, as in the first model, the unusual conformations for the backbone are all of low energy. It is tempting to suggest that these discontinuities in backbone conformations are ready-made biological recognition points, as discussed further elsewhere (Berman and Neidle, 1979).

All four models discussed in this section, however distinct in detail they may be,

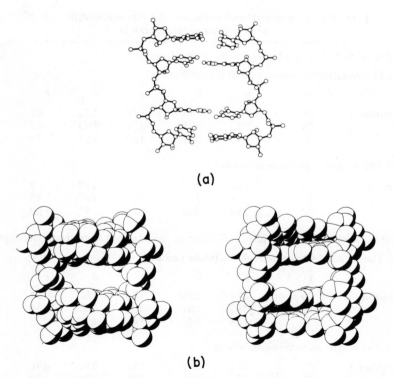

(a)

(b)

Figure 8.11   The neighbour exclusion site model (Berman and Neidle, 1979). (a), (b) Views as in figure 8.10.

share several common features. They all have non-colinear helical (or base-paired) residues either side of the intercalation site, and have adjacent residues affected in some measure by the changes in geometry at this site.

The success of the dinucleoside crystal structures in illuminating aspects of the intercalation problems has somewhat obscured the fact that the direct conformational relevance of these to polymeric situations has yet to be demonstrated. We do not know whether the dominant feature of glycosidic angle asymmetry persists in polynucleotide complexes. We are not, even at the dinucleoside level, yet able to understand phenomena such as sequence–preference binding. It is to be hoped that further structural studies of complexes in single crystals and fibres will illuminate both these and the more general questions posed at the onset of this chapter.

ACKNOWLEDGEMENTS

Support from the Cancer Research Campaign (Career Development Award), NATO, and the International Union Against Cancer, is gratefully acknowledged. The author is also grateful to A. Rich, H. M. Sobell and M. Sundaralingam for information in advance of publication, and H. M. Berman for much useful discussion.

REFERENCES

Alden, C. J. and Arnott, S. (1975). *Nucl. Acids Res.*, **2**, 1701.
Alden, C. J. and Arnott, S. (1977). *Nucl. Acids Res.*, **4**, 3855.
Arnott, S., Smith, P. J. C. and Chandrasekaran, R. (1976). In *Handbook of Biochemistry and Molecular Biology* (ed. G. D. Fasman), 3rd ed., vol. 2, section B, Chemical Rubber Co., Cleveland, Ohio, p. 411.
Berman, H. M. and Neidle, S. (1979). In *Stereodynamics of Molecular Systems* (ed. R. H. Sarma), Pergamon Press, New York, p. 367.
Berman, H. M., Neidle, S., Stallings, W., Taylor, G., Carrell, H. L., Glusker, J. P. and Achari, A. (1979). *Biopolymers*, **18**, 2405.
Berman, H. M., Neidle, S. and Stodola, R. K. (1978). *Proc. natn. Acad. Sci. U.S.A.*, **75**, 828.
Bond, P. J., Langridge, R., Jennette, K. W. and Lippard, S. J. (1975). *Proc. natn. Acad. Sci. U.S.A.*, **72**, 4825.
Cairns, J. (1962). *Cold Spring Harb. Symp. quant. Biol.*, **27**, 311.
Finkelstein, T. and Weinstein, I. B. (1967). *J. biol. Chem.*, **242**, 3763.
Fuller, W. and Waring, M. J. (1964). *Ber. Bunsenges Phys. Chem.*, **68**, 805.
Gale, E. F., Cundliffe, E., Reynolds, P. E., Richmond, M. H. and Waring, M. J. (1972). *The Molecular Basis of Antibiotic Action*, John Wiley, London.
Hamilton, L. D., Fuller, W. and Reich, E. (1963). *Nature, Lond.*, **198**, 538.
Jain, S. C., Bhandary, K. K. and Sobell, H. M. (1979). *J. molec. Biol.*, **135**, 813.
Jain, S. C. and Sobell, H. M. (1972). *J. molec. Biol.*, **68**, 1.
Jain, S. C., Tsai, C. C. and Sobell, H. M. (1977). *J. molec. Biol.*, **114**, 317.
Lerman, L. S. (1961). *J. molec. Biol.*, **3**, 18.
Lerman, L. S. (1963). *Proc. natn. Acad. Sci. U.S.A.*, **49**, 94.
Lerman, L. S. (1964a). *J. molec. Biol.*, **10**, 367.
Lerman, L. S. (1964b). *J. cell. comp. Physiol.*, **64**, 1.

Luzzati, V., Masson, F. and Lerman, L. S. (1961). *J. molec. Biol.,* **3**, 634.

Neidle, S. (1979). *Prog. Med. Chem.,* **16**, 151.

Neidle, S., Achari, A., Taylor, G. L., Berman, H. M., Carrell, H. L., Glusker, J. P. and Stallings, W. C. (1977). *Nature, Lond.,* **269**, 304.

Neidle, S., Taylor, G. L., Sanderson, M. R., Berman, H. M. and Shieh, H. S. (1978). *Nucl. Acids Res.,* **5**, 4417.

Neville, D. M. and Davies, D. R. (1966). *J. molec. Biol.,* **17**, 57.

Pigram, W. J., Fuller, W. and Hamilton, L. D. (1972). *Nature, Lond.,* **235**, 17.

Reddy, B. S., Seshadri, T. P., Sakore, T. D. and Sobell, H. M. (1979). *J. molec. Biol.,* **135**, 787.

Rosenberg, J. M., Seeman, N. C., Day, R. O. and Rich, A. (1976). *J. molec. Biol.,* **104**, 145.

Sakore, T. D., Jain, S. C., Tsai, C. C. and Sobell, H. M. (1977). *Proc. natn. Acad. Sci. U.S.A.,* **74**, 188.

Sakore, T. D., Reddy, B. S. and Sobell, H. M. (1979). *J. molec. Biol.,* **135**, 763.

Seeman, N. C., Day, R. O. and Rich, A. (1975). *Nature, Lond.,* **253**, 324.

Seeman, N. C., Rosenberg, J. M., Suddath, F. L., Kim, J. J. P. and Rich, A. (1976). *J. molec. Biol.,* **104**, 109.

Sheih, H. S., Berman, H. M., Dabrow, M. and Neidle, S. (1980). *Nucl. Acids Res.,* **8**, 85.

Smith, P. J. C. and Arnott, S. (1978). *Acta crystallogr.,* **A34**, 3.

Sobell, H. M. (1973). *Prog. Nucl. Acid Res. Mol. Biol.,* **13**, 153.

Sobell, H. M. and Jain, S. C. (1972). *J. molec. Biol.,* **68**, 21.

Sobell, H. M., Jain, S. C., Sakore, T. D. and Nordman, C. E. (1971). *Nature, Lond.,* **231**, 200.

Sobell, H. M., Lozansky, E. D. and Lessen, M. (1978). *Cold Spring Harb. Symp. quant. Biol.,* **43**, 11.

Sobell, H. M., Tsai, C. C., Jain, S. C. and Gilbert, S. G. (1977). *J. molec. Biol.,* **114**, 333.

Tsai, C. C., Jain, S. C. and Sobell, H. M. (1975a). *Proc. natn. Acad. Sci. U.S.A.,* **72**, 628.

Tsai, C. C., Jain, S. C. and Sobell, H. M. (1975b). *Phil. Trans R. Soc.,* **B272**, 137.

Tsai, C. C., Jain, S. C. and Sobell, H. M. (1977). *J. molec. Biol.,* **114**, 301.

Voet, D. and Rich, A. (1970). *Prog. Nucl. Acid Res. Mol. Biol.,* **10**, 183.

Wang, A. H. J., Nathans, J., van der Marel, G., van Boom, J. H. and Rich, A. (1978). *Nature, Lond.,* **276**, 471.

Wang, A. H. J., Quigley, G. J. and Rich, A. (1979). *Nucl. Acids Res.,* **6**, 3879.

Wang, J. C. (1974). *J. molec. Biol.,* **89**, 783.

Westhof, J. and Sundaralingam, M. (1980). *Proc. natn. Acad. Sci. U.S.A.,* **77**, 1852.

9

Oligonucleotide and polynucleotide – drug complexes in solution as investigated by NMR

Thomas R. Krugh

INTRODUCTION

This chapter discusses the solution complexes of drugs with oligo and polynucleotides. Although concentrating on magnetic resonance results, it also alludes to selected experiments using fluorescence, circular dichroism, and visible absorption spectroscopies, with a view to illustrating the advantages (and frequently the necessity) of combining the optical spectroscopic data with magnetic resonance data. In brief, not only does optical spectroscopy provide an independent means of obtaining kinetic, thermodynamic and equilibrium data, but it also gives an important correlation between the results for the binding of the drugs to oligonucleotides (that is, model systems) and polynucleotides (the systems of biological interest). The drugs that will be used as primary examples are actinomycin D, ethidium bromide, daunorubicin (or adriamycin) and 9-aminoacridine; the chemical structures of these molecules are shown in figure 9.1. Actinomycin D has been extensively studied over the last 20 years and, as shown later in this chapter, the interaction of actinomycin D with deoxyoligonucleotides is continuing to provide valuable information towards the goal of obtaining a molecular and thermodynamic basis for the binding of the drug to DNA. There are a number of reviews on the history, synthesis, binding studies and description of models for the binding of the actinomycins to DNA (for example, Hollstein, 1974; Lackner, 1975; Meienhofer and Atherton, 1977; Sobell, 1973; Wells, 1971; and the many references therein). Similarly, the reader is referred to the reviews by Henry (1976) and Arcamone *et al.* (1972) for background references on daunorubicin and adriamycin. Reinhardt and Krugh (1977, 1978) and Le Pecq (1971) provide background references on ethidium bromide. It has been well documented that the primary mode of binding of actinomycin D and ethidium bromide is by intercalation of the phenoxazone and the phenanthridinium

197

rings, respectively, into DNA. It is generally accepted that the planar portion of the daunorubicin molecule intercalates into DNA.

Both nuclear magnetic resonance and electron spin resonance have been used to study the complexes of intercalating drugs with nucleic acids. Since ^{1}H NMR has been the most extensively utilised, it has contributed the most to our knowledge of

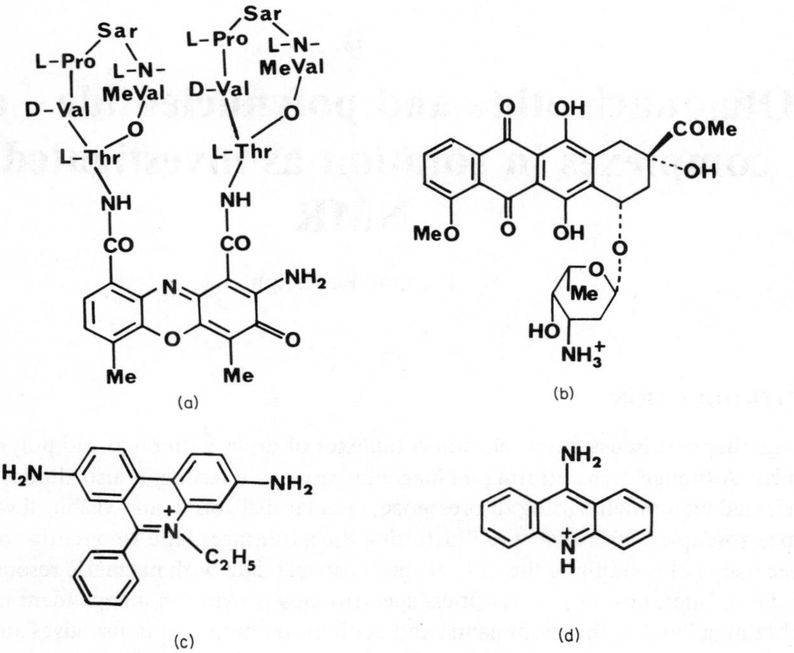

Figure 9.1 (a) Actinomycin D, (b) ethidium, (c) daunorubicin, (d) 9-aminoacridine.

drug–nucleic acid complexes to date. However, electron spin resonance experiments, using either paramagnetic metal ions as probes or spin labelled drug or nucleic acid molecules, have received increasing attention over the last few years (for example Chiao and Krugh, 1977; Dugas, 1977; Hong and Piette, 1976; Piette, 1974; Reuben and Gabbay, 1975; Sinha et al., 1976; Reuben, Baker and Kallenbach, 1976).

MAGNETIC RESONANCE PARAMETERS

The most useful parameters in nuclear magnetic resonance studies in solution are the chemical shifts, the indirect spin–spin coupling constants, the lineshape of the resonance (in those cases where chemical exchange phenomena are observed), and the relaxation times (especially the spin–lattice relaxation time, T_1). Each of these parameters may provide molecular or thermodynamic information on the drug–nucleic acid complexes as discussed below.

Chemical shifts

The chemical shifts of the drugs, the nucleic acids, and the drug–nucleic acid complexes are the parameters most frequently used to obtain information on the structure of the complexes. First, of course, the resonances of the drug and the nucleic acids used must be assigned. This can be a challenging task in itself, especially for complicated molecules like actinomycin D. The reader is referred to the reviews cited above and the articles cited below for a discussion of the methods used for the assignment of the resonances.

The changes in the chemical shifts of the drug and nucleic acid resonances that are observed upon drug–nucleic acid complex formation are important experimental parameters. The intercalated portion of the drug molecule is sandwiched between the aromatic base pairs and, therefore, the chemical shifts of the drug resonances are influenced by the ring current effects (see, for example, the texts by Dwek, 1973; Emsley, Feeny and Sutcliffe, 1965; James, 1975; and Ts'o, 1974a, b, for a discussion of ring current effects). With the measurement of the ring current shifts for several resonances, it is possible to construct molecular models for the drug–nucleic acid complex. Although the general agreement between the experimental data, x-ray crystallographic structures and the theoretical isoshielding contours has provided a firm basis for this type of analysis (for example, Borer, Kan and Ts'o, 1975; Giessner-Prettre et al., 1976; Kallenbach and Berman, 1977; Kearns, 1977; Krugh and Nuss, 1979), it is important to use as much data as possible when interpreting the changes in the chemical shifts. This is because there are several factors (in addition to the accuracy of the theoretical calculations) that should also be considered when trying to extract detailed geometrical information from the chemical shift changes ($\Delta\delta$) that accompany either drug–nucleic acid complex formation, or the helix to coil transition of nucleic acids (in either the presence or absence of added drugs). These complications may be divided into the major categories of solvent effects, hydrogen bonding effects, conformational averaging, self-aggregation of the drugs or nucleotides and 'other effects' (for example, local diamagnetic anisotropy and polarisation effects). The reader is referred to Borer, Kan and Ts'o (1975) and Krugh and Nuss (1979) for a detailed discussion of the importance and magnitudes of these complicating factors. Although the complexation shifts ($\Delta\delta$) of any nucleus (for example, ^{1}H, ^{19}F, ^{13}C, ^{31}P, ^{15}N, and so on) may be used to monitor complex formation, it is only the proton chemical shift data which have proved to be reliable in terms of ring current analyses. For example, it has been well established that ^{13}C chemical shifts are much more sensitive than proton chemical shifts to solvent and electronic effects (see, for example, Levy and Nelson, 1972; Stothers, 1972; Wehrli and Wirthlin, 1976). In addition, the inherently lower sensitivity in ^{13}C spectroscopy makes it more difficult to determine the infinite dilution chemical shifts that are needed to evaluate the induced chemical shifts ($\Delta\delta$ values) to be used in the ring current analysis.

Indirect spin–spin coupling constant analysis

In principle, the measurement and analysis of the proton–proton coupling constants of the deoxyribose protons, as well as the proton–phosphorus and the phosphorus-carbon coupling constants, will provide a great deal of stereochemical information on the structure of the oligonucleotides and the drug–oligonucleotide complexes. In practice, this is difficult information to extract even from proton spectra recorded at 360 MHz, because of severe overlap of the complicated resonance patterns from the many sugar residues. In addition, there may be substantial line-broadening present in the spectra due to slow rotation of the complexes. For example, an actinomycin D complex with a double-stranded deoxyhexanucleotide (one drug to one double helix) has a molecular weight in the vicinity of 5000, and thus appreciable dipolar linebroadening is anticipated, especially at low tempera-tures. Linebroadening may also result from chemical exchange effects. However, Lee and Tonoco (1978) and Patel and Shen (1978) have recently obtained useful information from an analysis of the coupling constants of the ribo and deoxyribo sugar protons of drug–nucleic acid complexes (as will be discussed later). A com-plete analysis of the proton–proton coupling constants of a drug–oligonucleotide complex may require the synthesis of specifically deuterated oligonucleotides (see, for example, Ezra *et al.*, 1977; Kondo and Danyluk, 1972; and references therein). This approach would overcome some of the difficulties encountered and allow for the determination of the important stereochemical changes that accompany drug-nucleic acid complex formation. The reader is referred to the texts by Ts'o (1974*a*, *b*) and the recent articles (and references therein) by Altona and Sundaralingam (1973), Borer, Kan and Ts'o (1975), Ezra *et al.* (1977) and Lee and Sarma (1976*a*, *b*) for more specific details on the interpretation of the coupling constants of the ribose and deoxyribose sugar protons in terms of the conformation of the molecule.

NUCLEAR RELAXATION EXPERIMENTS

The addition of paramagnetic metal ions or the use of spin-labelled drugs or nucleo-tides can provide valuable information on the structure of macromolecules through the measurement of the paramagnetic induced nuclear relaxation. The interaction between the unpaired electrons and the nuclei are usually dominated by a dipole-dipole interaction which has an r^{-6} dependence. The increase in the linewidths of the resonances and the simultaneous decrease in the spin-lattice relaxation times, T_1, thus provides a sensitive ruler to determine relative distances between the para-magnetic centre and various nuclei as first shown by Shulman and Sternlicht (1965) and Shulman, Sternlicht and Wyluda (1965). A number of workers have studied the binding of paramagnetic metal ions to nucleoside mono, di, and triphosphates as well as DNA (for example, see Anderson *et al.*, 1971; Chiao and Krugh, 1977; Cohn and Hughes, 1962; Reed, Leigh and Pearson, 1971; Reuben and Gabbay, 1975; Shulman, Sternlicht and Wyluda, 1965; and the many references therein). The reviews by Dwek (1973), Krugh (1976), Mildvan and Cohn (1970) and Swift (1973) provide the necessary background to the theory, control experiments, and illustrative examples of the use of paramagnetic induced nuclear relaxation. Since there have only been limited applications of this approach in the study of drug-nucleic acid complexes, the background material will not be discussed in the present chapter. A recent example of this approach is the work of Chiao and Krugh (1977) in which they exploited the preferential binding of manganese(II) ions to the 5'-terminal phosphate groups of mono and dinucleotides to measure the preferential binding of adenine and guanine mono and dinucleotides to the two nucleotide binding sites of actinomycin D. The change in the linewidth (which is proportional to the paramagnetic-induced nuclear relaxation) of the 4-CH_3, the 6-CH_3, and the H(7) and H(8) protons of the phenoxazone ring of actinomycin D were measured as Mn(II) was added to various solutions of actinomycin D in which only one of the nucleotides had a 5' (terminal) phosphate group. The preferential broadenings of the actinomycin D resonances were consistent with the ring current analysis of the chemical shift data for similar systems (Krugh, Mooberry and Chiao, 1977).

It has been already discussed that ^{13}C chemical shifts of drug–nucleic acid complexes are not likely to provide definitive structural information on the nature of the complex. However, the use of a paramagnetic probe bound selectively to the terminal phosphate group of a nucleotide (Chiao and Krugh, 1977) and the measurement of the paramagnetic-induced nuclear relaxation of the ^{13}C resonances of the drug may be an ideal way to obtain quantitative structural information. Since the chromophores of the drugs always contain a number of carbon nuclei, it should be possible to measure a number of relative distances and thus determine a detailed molecular structure. It will be important to determine if the selectivity of the binding of Mn(II) ions to the terminal phosphate group (which carries a −2 charge at neutral pH) is maintained in the oligonucleotide systems, or alternatively stated, the location of the Mn(II) binding site must be known. The same approach can be used with other nuclei (for example ^{1}H NMR), but a number of important drugs, such as daunorubicin or adriamycin, have chromophores with only a limited number of protons. For these drugs the paramagnetic induced nuclear relaxation experiments (using Mn(II) or other relaxation probes) may be the only way of providing detailed structural information.

Reuben, Adawadkar and Gabbay (1976) have also used Mn(II)-induced ^{1}H nuclear relaxation experiments to determine the orientation of the 3,8-dimethyl-N-methyl-phenanthrolinium cation when it is intercalated into DNA. The addition of DNA to a solution of the 3,8-dimethyl-N-phenanthrolinium had a negligible effect on the chemical shifts and spin-lattice relaxation rates. The subsequent addition of Mn(II) ion resulted in a marked difference in the paramagnetic induced nuclear relaxation of the various protons (as much as a hundredfold difference) from which Reuben et al. concluded that in the intercalation complex the long axis of the 3,8-dimethyl-N-methyl-phenanthrolinium cation is almost perpendicular to the hydrogen bonds of the DNA base pairs, in agreement with previous physical and spectral data for the same system (Gabbay, Scofield and Baxter, 1973). Although this is an interesting approach, we comment that it will not be a generally useful one to the study of drug–DNA complexes because the intercalation of compounds into DNA results in a severe broadening of the resonances (see, for example, Gabbay, Scofield and Baxter (1973)). Reuben, Adawadkar and Gabbay (1976) circumvented this problem by using a solution that contained a large excess of unbound cation (0.05 per cent bound). The spectral parameters were a weighted average of the parameters for the bound and free phenanthrolinium cations, since the mean residence time in the bound state ($< 3 \times 10^{-5}$ s) is much smaller than the relaxation rates (and the $\Delta\delta$ values) for the various protons of the DNA bound molecule. Unfortunately, many important drugs dissociate relatively slowly from DNA, which vitiates the usefulness of paramagnetic-induced nuclear relaxation experiments with DNA. Actinomycin D is an extreme example of a slowly-dissociating drug in that the dissociation time constants range from 10 to 1500 s (Krugh *et al.*, 1979; Müller and Crothers, 1968). Another complicating problem is that the large excess of unbound cations or drugs may result in aggregation of the positively charged molecules along the outside of the helix near the negatively charged phosphate groups (a well-known phenomenon for other positively charged intercalating chromophores such as the acridines or ethidium bromide). If the outside bound molecules contributed to the observed magnitude of the paramagnetic induced nuclear relaxation, then neglecting this contribution may lead to errors in the interpretation of the data. However, paramagnetic-induced nuclear relaxation experiments can be a valuable tool in the study of drug–nucleic acid complexes, especially in the use of oligonucleotide model systems.

HISTORICAL INTRODUCTION TO NMR STUDIES OF DRUG–NUCLEIC ACID COMPLEXES

Before outlining the detailed studies on actinomycin D, ethidium bromide and 9-aminoacridine it is appropriate to provide a brief background and selected review of the many papers that have used nuclear magnetic resonance techniques in the study of drug–nucleic acid complexes or, more generally, small molecule–nucleic acid complex formation. In 1966 Chan, Bangerter and Peter published the first in a series of papers in which proton magnetic resonance was used to show that purine formed intercalated complexes with a series of (single-stranded) dinucleoside monophosphates (for example, Chan, Bangerter and Peter, 1966; Chan and Kreishman, 1970). In 1971, Kreishman, Chan and Bauer published a detailed study in which proton magnetic resonance was used to show that in aqueous solutions ethidium bromide self-aggregates by the stacking of the phenanthridinium rings. Ethidium also formed mutually stacked aggregates with several uracil residues. An analysis of the chemical shift changes and preferential linebroadening of the H(1) and H(10) proton resonances that were observed upon complex formation with UpU provided evidence that the phenanthridinium ring of ethidium was sandwiched between the adjacent uracil bases with the H(1) and H(10) protons directed towards the ribose-phosphate backbone of the dinucleoside monophosphate. The proton magnetic resonance data of ethidium complexes with poly U were less conclusive than the ethidium-dinucleotide data, but the preferential broadening of the H(1) and H(10) proton resonances again indicated the intercalation of ethidium between adjacent uracil bases of poly U. It is noteworthy that the Kreishman, Chan and Bauer (1971) paper is the first report of the use of nuclear magnetic resonance to study drug–nucleic acid complexes at the dinucleoside level. Several other groups had used proton magnetic resonance to study a variety of complexes formed between aromatic chromophores with mononucleosides or mononucleotides, and obtained the general result that stacked aggregates are formed. However, except for actinomycin D, these studies have not been definitive in providing a molecular interpretation of the interaction of the drugs with double-stranded DNA or RNA because the mononucleotides cannot form a base-paired duplex.

Krugh and Neely (1973b) have studied the nuclear magnetic resonance spectra of a series of actinomycin D complexes with both complementary and noncomplementary deoxydinucleotides. Patel (1974a, b) used ^{1}H and ^{31}P spectroscopy to investigate actinomycin D complexes with pdG-dC and the self-complementary deoxyhexanucleotide d-ApTpGpCpApT. Patel and coworkers have performed extensive nuclear magnetic resonance studies on the binding of actinomycin D to deoxyoligonucleotides as well as the binding of several other drugs to deoxyoligonucleotides and synthetic polynucleotides (Patel, 1976a, 1977b; Patel and Canuel, 1976, 1977a, b, c; Patel and Shen, 1978). Davidson et $al.$ (1977) have recently reported ^{1}H and ^{13}C NMR and ultraviolet visible spectroscopic studies on the binding of a methylated quinacrine derivative and propidium diiodide to dinucleoside monophosphates.

Heller, Tu and Maciel (1974) studied the interaction of poly A_{20-24} with poly U_{20-24} to form the poly (a) ·poly (U) duplex (20–24 duplex nucleotides in length). The binding of miracil D [1-(2-diethylaminoethylamino)-4-methyl-10-thiaxanthenone] to the poly (A) ·poly (U) (20–24) duplex and the constituent single-stranded oligomers was also studied by proton magnetic resonance. A preferential broadening and a 0.3 ppm upfield shift of the 4-methyl resonance on the thiaxanthenone ring was in contrast to the unshifted resonances for the two terminal methyl resonances of the diethylaminoethylamino chain. These data are consistent with the intercalation of miracil D into the double-stranded poly (A) ·poly (U) duplex as had been proposed from other studies (see Heller, Tu and Maciel, 1974). Chien, Grollman and Horwitz (1977) have studied the interaction of bleomycin A_2 and tripeptides with DNA. A number of compounds have recently been synthesised that contain two intercalating chromophores connected by various chemical 'chains'. The acridine dimers were the first synthetic bisintercalators (Barbet, Roques and Le Pecq, 1975) and proton magnetic resonance has been used to study the conformation and interactions of an acridine dimer with mono and dinucleotides (Barbet et $al.$, 1976).

Gabbay and coworkers have used proton magnetic resonance and a variety of other techniques to investigate the interaction of a large number of synthetic oligopeptides with nucleic acids (see, for example, Gabbay, Scofield and Baxter, 1973; Gabbay et $al.$, 1976 and the many references therein). These extensive, interesting studies have provided a great deal of stereochemical and topological information on oligopeptide–nucleic acid interactions, especially the partial intercalation of aromatic amino acids. Dimicoli and Helene (1974) have also studied the interactions of aromatic residues of proteins with nucleic acids.

Reuben, Baker and Kallenbach (1978) have recently studied the proton NMR spectra of 9-aminoacridine and complexes of 9-aminoacridine with dG-dC, dC-dG, and the hexanucleotide d-ApTpGpCpApT. 9-Aminoacridine is a particularly inter-

(a)   (b)

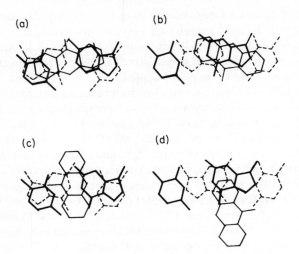

(c)   (d)

Figure 9.2 Scale drawing of four possible complexes of dinucleotide with 9-aminoacridine: (a), (b) based on the crystal structures of Sakore *et al.* (1977); (c), (d) reproduced from Reuben *et al.* (1978).

esting molecule because Sobell and coworkers (Sakore *et al.*, 1977) have reported the x-ray crystallographic structure of a 2:2 complex of 9-aminoacridine with i^5-CpG. This cocrystalline complex contains two different kinds of intercalated 9-aminoacridines. The first of these involves a pseudosymmetric stacking interaction of the 9-aminoacridine between guanine-cytosine base pairs. The second configuration is an asymmetrically intercalated 9-aminoacridine. A schematic illustration of the two types of intercalative structures observed in the crystal structure is shown in figure 9.2a, b. Figures 9.2c, d represent two additional classes of trial and configurations used by Reuben, Baker and Kallenbach (1978) to make a comparison between the experimental and calculated values of the induced shifts. The ^{1}H NMR titration data showed that two dG-dC molecules would form a complex with 9-aminoacridine whereas the titration data for d-ApTpGpCpApT suggested that three 9-aminoacridines bind to each hexanucleotide duplex, although the chemical shifts of the 9-aminoacridine resonances continued to change as additional hexanucleotide was added (beyond the apparent minimum stoichiometric amount) so the possibility of the stacking of the drugs at the end of the helix must also be considered.

Theoretical induced shifts were calculated on the basis of the overlap of the 9-aminoacridine with the adjacent C·G or A·T base pairs as shown in figure 9.2. Reuben, Baker and Kallenbach (1978) concluded that either configuration (b) or configuration (d) (figure 9.2) provided good agreement between the experimental and calculated values of the induced shifts. The chemical shift data appear to rule out the pseudosymmetric intercalated structure as the predominant complex formed in solution (figure 9.2a), which contrasts with the proposal of Sobell

(Sakore *et al.*, 1977) that this is probably the type of complex expected for DNA intercalation. The agreement between the experimental and calculated Δδ values for both structures (b) and (d) should serve as a cautionary note to those interpreting chemical shift data in drug–nucleic acid complexes. (Structure (d) results in the same average chemical shifts as (b) because the rapid chemical exchange produces an averaging of the very large upfield shifts of the 9-aminoacridine resonances that are located between the base pairs and the negligible induced shifts of the 9-amino-acridine resonances on the end of the chromophore which is pointing away from the base pairs.)

Actinomycin D complexes with deoxyoligonucleotides

The reader is referred to the reviews by Krugh and Nuss (1979) and Meienhofer and Atherton (1977) for an introduction to the extensive literature on the actinomycins. The intercalation model for the actinomycin–DNA complex was proposed by Müller and Crothers (1968) and received support from the hydrodynamic experiments of Waring (1970) and Wang (1971). Actinomycin has a general requirement for the presence of guanine at the intercalation site and in 1970 Arison and Hoogsteen published a detailed PMR study on actinomycin D and a 2:1 dGMP: actinomycin D complex. The following year Sobell *et al.* (1971) reported the

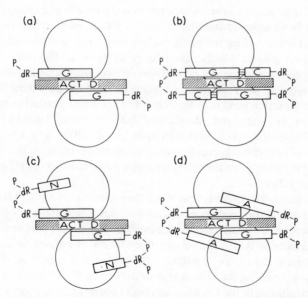

Figure 9.3   Schematic illustrations of actinomycin D complexes. The phenoxazone ring projects out of the plane of the paper and is indicated by the dashed lines. The cyclic pentapeptide rings are schematically illustrated as circles. (a) Actinomycin D-deoxyguanosine 5′-monophosphate complex; (b) actinomycin D-pdG-dC complex, illustrating the formation of an intercalated complex; (c) actinomycin D-pdN-dG complex; (d) actinomycin D-pdG-dA complex, illustrating steric interference of the noncomplementary bases. (From Krugh and Neely, 1973*b*.)

crystal structure of a 2:1 deoxyguanosine complex with actinomycin D and from this structure proposed a detailed stereochemical model for the actinomycin *D*-DNA complex (Sobell and Jain, 1972). Krugh (1972) monitored the change in the absorbance of actinomycin D (at 425 nm) as a function of the concentration of added deoxydinucleotides and from this data proposed that the complexes formed were of the types schematically illustrated in figure 9.3. The highly cooperative binding of pdG-dC was evidence that two dinucleotides were binding to actinomycin D (figure 9.3b). The co-operativity presumably resulted from the formation of the six hydrogen bonds associated with the two G·C Watson-Crick base pairs. An intriguing aspect of the optical titration data is that the self-complementary deoxydinucleotide pdC-dG formed a 2:1 stacked complex (figure 9.3c) instead of an intercalated complex in which the actinomycin D chromophore would be intercalated at a dC-dG sequence. This suggests that both stacking and hydrogen bonding forces are important in stabilising the structures of the dinucleotide complex. By monitoring several of the actinomycin D resonances in the proton NMR spectra of this series of dinucleotide complexes, Krugh and Neely (1973*a*, *b*) were able to verify that in the presence of excess dinucleotides ($> 2:1$ dinucleotide:drug ratios) the predominant complexes formed are those shown in figure 9.3. For example, the stacking patterns observed for the N nucleotides in the four pdN-dG deoxydinucleotide complexes with actinomycin D were verified (in part) by monitoring the proton resonances of the two methylvaline N—CH_3 resonances located on the face of the pentapeptides adjacent to the chromophore and, therefore, in a location that would be particularly sensitive to the ring current of the nucleotide base one removed from the base that is stacked on the phenoxazone ring. There are two other noteworthy aspects of these spectra. Firstly, the induced shifts of the H(7) and H(8) resonances of the phenoxazone ring in the pdN-dG complexes with actinomycin D (figure 9.3c) are consistent with the ring currents calculated from stacking the guanine base in an orientation similar to that observed for deoxyguanosine in the crystal structure. On the other hand, the induced shifts of the H(7) and H(8) resonances in the pdG-dN complexes (0.63 and 0.16 ppm for pdG-dC) contain, in addition to the ring currents, local diamagnetic anisotropic contributions from the sugar phosphate backbone that result from the proximity of these groups to the H7 and H8 resonances (figure 9.3b).

Another important point is that all of these spectra are subject to conformational averaging and thus the geometries deduced reflect this averaging. Fortunately, the cyclic pentapeptides greatly reduce the number of possible conformations that are allowed. Consequently, for the cooperative binding of pdG-dC to actinomycin D one cannot form the type of complexes illustrated in figures 9.2c, d for 9-aminoacridine. By comparing the chemical shift data for pdG-dC complex formation and the chemical shift data for the actinomycin D complex formation with the other dinucleotides, Krugh and Neely (1973*b*) concluded that the predominant complexes formed are those which are illustrated in figure 9.3.

Patel (1974*a*, *b*) extended the NMR study of actinomycin D complexes to the hexanucleotide level by measuring ^{31}P and ^1H NMR spectra of a 2:1 (nucleotide strand:actinomycin D) complexes with d·(ApTpGpCpApT) and d·(pGpC)$_3$. Large downfield shifted resonances (1.5 ppm and 2.5 ppm) were observed for the two phosphorus resonances at the intercalation site. Unfortunately, it is not possible to interpret the phosphorus-31 chemical shift data in terms of detailed changes in the structure of the nucleic acid backbone, although efforts in this area are still continuing (see, for example, Gorenstein, 1975; Gorenstein and Kar, 1975; Gorenstein and Luxon, 1979; Gorenstein *et al.*, 1976; Mariam and Wilson, 1979; Patel, 1976*a*, *b*, 1977*a*; Reinhardt and Krugh, 1977; Salemink, Swarthof and Hilbers, 1979). The conformational changes in the sugar-phosphate backbone that accompany intercalaction undoubtedly have important biological ramifications and it is unfortunate that ^{31}P NMR chemical shifts do not presently provide quantitative structural information. Qualitatively, however, the ^{31}P NMR data are informative in that a comparison of the data for the formation of different drug–oligonucleotide complexes (for example, Reinhardt and Krugh, 1977) shows that the conformation of the backbone is dependent upon the sequence of the nucleotides at an actinomycin site and upon the nature of the chromophore which intercalates. The conformational differences in the backbone structure have also been observed in recent crystal structures as discussed in chapter 8. Krugh and coworkers (for example, Krugh *et al.*, 1979) have shown that the dissociation rate of actinomycin D from synthetic DNA polymers is very dependent upon the sequence at the intercalation site. For instance, under essentially identical conditions, the dissociation lifetime of an actinomycin D-poly(dG-dC)·poly(dG-dC) complex is approximately two orders of magnitude larger than the dissociation lifetime for an actinomycin D-poly(dG-dT)· poly(dG-dT) complex. In view of the ^{31}P NMR data of Reinhardt and Krugh (1977) it is interesting to speculate that the higher activation energy for the dissociation of actinomycin D from poly(dG-dC)·poly(dG-dC) is intimately associated with a different conformation of the complex at the intercalation site. The different dissociation lifetimes have been correlated to a sequence-dependent physiological activity of the actinomycin (Krugh and coworkers, unpublished observations) and the next important link is to interpret the phenomenon in terms of a molecular geometry for the complexes.

Recording the ^{1}H NMR spectra of the drug–oligonucleotide complexes in H_2O solution allows for the observation of the NH_2 and N—H—N resonances which can provide valuable information on hydrogen bonding as well as information on the stability of the drug–nucleic acid complexes and the oligonucleotides (for example, see Krugh and Reinhardt, 1975; Patel 1974*a*, *b*, 1979). These experiments typically require the availability of a high frequency ($\geqslant$ 270 MHz) NMR spectrometer, although techniques have been developed which also allow for the measurements of spectra on lower frequency instruments (see the discussion in Krugh and Schaefer, 1975). When complexed with d-(ApTpGpCpApT) actinomycin D intercalates into the (dG-dC) sequence of the hexanucleotide duplex, and the proton resonances of the hydrogen-bonded resonances were used by Patel (1974*b*) as a probe of the distortion in the duplex that results from complex formation. Although it is reasonably straightforward to observe and measure the chemical shifts of the N—H—N resonances (10–14 ppm range), it is much more difficult to unequivocally assign these resonances to individual G·C or A·T base pairs in the drug–oligonucleotide complexes. In addition, it is not easy to interpret these changes which accompany intercalation in terms of structural perturbations of the oligonucleotide duplex. However, during the last few years a number of advances have been made which facilitate the interpretation of this data.

Ethidium bromide

The binding of ethidium bromide to DNA has been well characterised by fluorescence, visible absorption, and circular dichroism studies (Le Pecq, 1971; Le Pecq, Yot and Paoletti, 1964; Reinhardt and Krugh, 1978; Waring, 1965; and references therein). Prior to the experiments of Krugh and coworkers (Krugh, 1974; Krugh

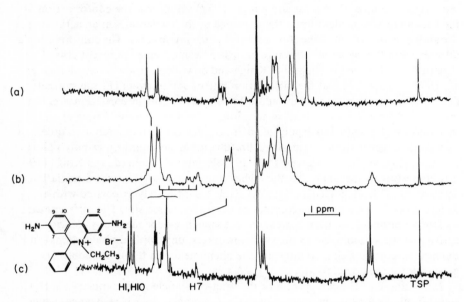

Figure 9.4   100-MHz proton Fourier transform NMR spectra of (a) 0.9 mM CpG, (b) 2.3 mM CpG + 0.75 mM ethidium bromide and (c) 0.75 mM ethidium bromide. The spectra were run at 25 °C with a 180-τ-90 (accumulate) pulse sequence to minimise the residual HDO peak. The use of this pulse sequence accounts for the small H7 peak since it has a longer T_1 value than the other EB ring protons. The movement of the peaks has also been followed by the incremental addition of the CpG. (From Krugh, Wittlin and Kramer, 1975.)

and Reinhardt, 1975; Krugh, Wittlin and Cramer, 1975) it was generally assumed that ethidium bromide had equal affinity for all intercalation sites on DNA. However, it has now been well established that ethidium exhibits a range of affinity constants for the various intercalation sites on both DNA and RNA. Nuclear magnetic resonances studies have been performed on complexes of ethidium with both deoxy and ribodinucleotides (Krugh, 1974; Krugh and Reinhardt, 1975;

Table 9.1 Upfield chemical shifts of ethidium in the presence
of oligonucleotides[a, b]

| Proton | δ (free ethidium) − δ (ethidium in complex) in ppm | | | |
	CpG[c]	CpUpG[d]	GpUpG[e]	GpUpG + CpC[f]
H1	0.58	0.52	0.24	0.45
H2	0.92	0.66	0.24	0.68
H4	1.22	0.75	0.21	0.61
H7	1.05	0.63	0.28	0.62
H9	0.85	0.66	0.22	0.56
H10	0.58	0.49	0.25	0.38

[a]Shifts are reported at 10 °C in D_2O solution containing EDTA.
[b]Spectra were recorded on a Bruker HXS 360-MHz spectrometer (Stanford Magnetic Resonance Laboratory).
[c]5 mM CpG + 2.5 mM ethidium.
[d]3.2 mM CpUpG + 1.6 mM ethidium.
[e]2 mM GpUpG + 2 mM ethidium.
[f]2 mM GpUpG + 2 mM CpC + 2 mM ethidium. At these concentrations the complex is not fully formed. (From Lee and Tinoco, 1978.)

Reinhardt and Krugh, 1977) as well as oligonucleotides and polynucleotides (Patel and Canuel, 1976, 1977a). All of these experiments show conclusively that ethidium bromide intercalates, as illustrated by the proton NMR data in figure 9.4 and table 9.1. The spectra in figure 9.4 were recorded at 100 MHz and are from Krugh *et al.* (1975), while the data in the table were recorded at 360 MHz and are from Lee and Tinoco (1978). The availability of the 360 MHz instrument allowed Lee and Tinoco to follow all of the resonances during the titration. The large $\Delta\delta$ values observed by Krugh and Reinhardt (1975), Patel and Canuel (1976, 1977a) and Lee and Tinoco (1978) for the ethidium complexes (table 9.1) lead to the conclusion that ethidium bromide forms an intercalated complex with both CpG and dCpdG. The unusual types of complexes that were considered as possibilities for 9-aminoacridine (figure 9.2c, d) are not consistent with the large ring current shifts observed for all six phenanthridinium ring protons in the ethidium complexes with CpG (as well as other pyrimidine (3′-5′) purine dinucleotides). In addition, the presence of the N-ethyl group and the phenyl ring attached to the phenanthridinium ring restrict the possible geometries for the complex.

Ethidium-oligonucleotide complexes with a bulged base

Lee and Tinoco (1978) studied the binding of ethidium bromide to CpG, CpUpG, and an equimolar mixture of GpUpG + CpC. The changes in the proton chemical shifts of ethidium upon complex formation were interpreted in terms of the formation of the complexes shown in figure 9.5. The interesting proposal of the formation of bulged-based structures (figure 9.5) suggests that ethidium may cause frameshift mutations by intercalation into transient bulges formed during replica-

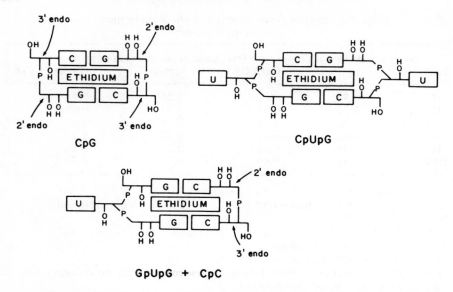

Figure 9.5 Proposed structures in aqueous solution for the complexes containing ethidium. In these complexes the 3′ linked sugar of each dinucleoside phosphate tends to the 3′ *endo* conformation, whereas the 5′ linked sugar tends to the 2′ *endo* conformation. (From Lee and Tinoco, 1978.)

tion, recombination, or repair. It will be interesting to study the binding of ethidium to longer oligonucleotides where the double-stranded form has a pre-existing defect due to mismatched base pairing.

SEQUENCE PREFERENCES AND BISINTERCALATORS

The extensive oligonucleotide binding studies cited above have clearly shown that both actinomycin D and ethidium bromide preferentially intercalate at certain sequences on DNA. It is important to note, however, that these are not absolute sequence specificities but, rather, a range of intrinsic affinity constants are anticipated for the various intercalation sites on DNA (and RNA). There is no question that some dyes and drugs exhibit preferential binding to base pairs or base sequences (see, for example, Krugh and Nuss, 1979; Müller and Crothers, 1975; Wartell, Larson and Wells, 1974; and the references therein). The presence of sequence recognition and preferential binding creates difficulties in the binding analysis, but it also creates exciting possibilities for the design of new drugs (such as *bis*-intercalators) which may be as sequence-specific as native proteins. Although, these possibilities have yet to be realised, the synthesis and study of *bis*-intercalators is being pursued by several groups, in particular the natural antibiotic echinomycin (Waring and Wakelin, 1974; Wakelin and Waring, 1976), and the synthetic *bis*-intercalators (Barbet, Rogues and Le Pecq, 1975, 1976; Canellakis *et al.*, 1976; Capelle *et al.*, 1979; Dervan and Becker, 1978; Kuhlman, Charbeneau and Mosher, 1978; and references therein).

GEOMETRY OF THE HELIX AT THE INTERCALATION SITE

Neidle (chapter 8) has reviewed the structures of the various cocrystalline complexes which have been determined by x-ray crystallography, and has discussed the question of the conformation of the sugar pucker at the intercalation site. Sobell and coworkers (Sobell *et al.*, 1977, 1980) have proposed that the mixed sugar puckering observed in most of the crystalline complexes is a common feature of intercalation whereas Berman, Neidle and Stodola, 1978, and Neidle (chapter 8) conclude that there is a continuum of stereochemically plausible 'opened-up' dinucleoside phosphate structures and that the geometry would thus depend upon the nature of the intercalating drug. It is important to consider the solution complexes in this context because it is quite possible that several conformations will be present in solution since the process of crystallisation could result in the selection of only one of these nearly energetically equivalent conformations.

Lee and Tinoco (1978) and Patel and Shen (1978) have recently utilised 360 MHz NMR spectroscopy to determine the conformation of the sugar rings in complexes of dinucleosides with ethidium bromide and propidium diiodide, respectively. Lee and Tinoco (1978) observed that at 40 °C, the ribose sugar of unbound CpG was 62 per cent C3′ *endo* for Cp and 52 per cent C3′ *endo* for pG; in the presence of ethidium, such that about half the dinucleoside phosphate was complexed, the $J_{1,2}$ values indicate that the 3′ sugar (Cp) tends to increase the proportion in a 3′ *endo* conformation while the 5′ sugar (pG) tends to increase the population of the 2′*endo* conformer. These data support the model building that incorporates the mixed sugar puckering observed in the solid state complex of ethidium with iodo-CpG and iodo-UpA (Tsai *et al.*, 1975; Jain *et al.*, 1977). The values for the percentage of riboses in each of the two conformations were not given for the ethidium: CpG complex in this initial communication. It should be noted that Lee and Tinoco (1978) performed their experiments in the presence of excess nucleoside, and thus they would have to extrapolate their results (by a factor of two) to obtain estimates for the sugar conformation in the complex. Lee and Tinoco (1978) also state that ethidium complexes with deoxyribodinucleoside monophosphates show similar behaviour; that is, the sugars tend toward a C3′ *endo* (3′-5′)C2′ *endo* conformation. Interestingly enough, they also state that an ethidium complex with hybrid helices (one strand ribose–one strand deoxyribose) also shows similar behaviour.

Patel and Shen (1978) monitored the temperature dependence of the chemical shifts and coupling constants of propidium diiodide solutions with CpG and with dCpdG. The chemical shift data show that propidium forms an intercalated complex with the self-complementary dinucleotides (both ribo and deoxyribo). The temperature dependence of the $J_{1'2'}$ coupling constants for the CpG:propidium solutions are shown in figure 9.6a. Because of linebroadening and spectral overlap, Patel and Shen were able to measure the coupling constants only at temperatures above 50 °C (figure 9.6). At ~50 °C Patel and Shen estimated that ~55 per cent of the propidium molecules are complexed with the dinucleoside monophosphates. Thus an extrapolation is required in order to determine the conformation of the sugar residues in the fully complexed form. For the propidium experiments with

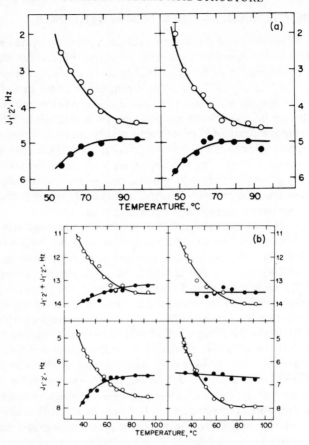

Figure 9.6   (a) The temperature dependence of the vicinal $J_{1'2'}$ coupling constant in the PrdI$_2$·C—G complex (Nuc/D = 4) in the absence (left) and presence (right) of 1 M NaCl in 0.1 M phosphate/1 mM EDTA, pH 6.55, in ^{2}H$_2$O. The cytidine and guanosine coupling constants are designated by ○ and ●, respectively. (b) Temperature dependence of the vicinal $J_{1'2'} + J_{1'2''}$ coupling constant sum and the vicinal $J_{1'2'}$ coupling constant in the PrdI$_2$·dC· dG complex in the absence (Nuc/D = 2.8) (left) and presence (Nuc/D = 3.2) (right) of 1 M NaCl in 0.1 M phosphate/1 mM EDTA, pH 6.55, in ^{2}H$_2$O. (From Patel and Shen, 1978.)

CpG we note that the largest change is observed for the $J_{1'2'}$ coupling constant of the cytidine sugar (figure 9.6) which shows that the Cp residue favours the C3′ *endo* conformation in the complex. If we try to quantitate Patel and Shen's data (figure 9.6) by using the relationship per cent C3′ *endo* = 10(10-$J_{1'2'}$) (Lee and Sarma, 1976a, b) then we estimate that Cp is 55 per cent C3′ *endo* at 95 per cent, while at 50° (when the fraction of the drug complexed is ∼55 per cent) approximately 75 per cent of the Cp sugars are in the C3′ *endo* conformation. Although we may linearly extrapolate these values to 100 per cent complexation it should be obvious that this extrapolation requires important assumptions. The most critical assumption concerns the type of complexes that exist at 50° since this temperature is approximately equal to the transition midpoint in the melting curve (for example, the fraction of the propidium in a complex at ~50° is ~0.55). It is not known what fractions of the 'complexes' have a 2:1 CpG:propidium ratio as opposed to a 1:1 complex. In other words, is the data from 95° to 50° a reliable predictor of the low temperature (20 °C) complex in which propidium is intercalated between the two C·G base pairs? Having stated the cautionary note, and having no other data to suggest otherwise, we will assume that the data is a reliable predictor of the stable solution complex. Extrapolating these values to 100 per cent complexation suggests that more than 90 per cent of the Cp residues will be in the C3′ endo conformation in the propidium:CpG complex. On the other hand, at 95 °C the pG residue of CpG has an approximately 50:50 mixture of the C3′ endo and C2′ endo sugar conformations which changes to ~43 per cent C3′ endo when the temperature is lowered to 50 °C, where the fraction of the drug in a complex is ~0.55. An approximate extrapolation to 100 per cent complexation suggests that 35–40 per cent of the pG residues will be in the C3′ endo conformation (and consequently 60–65 per cent of the pG residues will be in the C2′ endo conformation). Thus, the data of Patel and Shen (1978) show that the C3′ endo-(3′, 5′)-C2′ endo conformation is favoured in the propidium diiodide complex with CpG, but the extrapolated data also suggest that a substantial fraction (~30–40 per cent) of the dinucleosides will have the C3′ endo-(3′, 5′)-C3′ endo conformation in the complex. If this approximate extrapolation is appropriate, then the solution data indicate that the 5′ residue of CpG has only a relatively small energy difference between the two different sugar puckers in the intercalated complex with propidium (0.4–0.7 kcal/mole). We reach the same conclusion concerning the small energy differences between the two sugar puckers from an analysis of Patel and Shen's data for the dC-dG complex with propidium (figure 9.6b). We also note that Patel and Shen (1978) observed that in the presence of 1 M NaCl the pdG residue does not exhibit any change in the coupling constant upon complex formation. While the effect of salt is not understood at the present time, the observation of the phenomenon is interesting and suggests that caution must be exercised in the interpretation of the data.

In summary, we agree with Patel and Shen (1978) and Lee and Tinoco (1978) when they state that their data indicate that the mixed sugar pucker is favoured in the intercalated complexes studied; equally important may be our observation that Patel and Shen's data suggest that the mixed sugar pucker is not a highly favoured conformation in the propidium complexes with the deoxy and ribodinucleoside monophosphates. The resolution of this question and the role of sugar pucker in solution complexes has yet to be finally determined. A number of additional experiments on the solution complexes and crystal structures of several classes of drugs with oligonucleotides (for example tetra or hexanucleotides) are required in order to obtain more definitive answers to these interesting questions.

CONCLUDING REMARKS

Nuclear magnetic resonance spectroscopy has been used to elucidate the structure, kinetic and thermodynamic properties of drug–nucleic acid complexes. In the study of nucleic acids and drug–nucleic acid complexes, x-ray crystallography and nuclear magnetic resonance spectroscopy (in combination with optical and kinetic studies) have developed an important synergistic relationship which, hopefully, will grow and provide continuing progress toward the goal of understanding the molecular basis of drug action.

ACKNOWLEDGMENTS

The authors' research discussed in this chapter, and the writing of this review, have been supported by research grants (CA-14103 and CA-17865), a Career Development Award (CA-00257) from the National Cancer Institute, DHEW, and an Alfred P. Sloan Fellowship. Portions of this chapter have been reproduced, with permission, from a chapter by T. R. Krugh and M. E. Nuss in *Biological Applications of Magnetic Resonance* (ed. R. E. Shulman) Academic Press, New York, 1979 (copyright Bell Telephone Laboratories, Inc.).

REFERENCES

Altona, C. and Sundaralingam, M. (1973). *J. Am. chem. Soc.*, **95**, 2333.

Anderson, J. A., Kuntz, G. P. P., Evans, H. H. and Swift, T. J. (1971). *Biochemistry*, **10**, 4368.

Arcamone, F., Cassinelli, G., Franceschi, G., Penco, S., Pol, C., Redaelli, S. and Selva, A. (1972). In *International Symposium on Adriamycin* (ed. S. K. Carter et al.), Springer-Verlag, New York, p. 9.

Arison, B. H. and Hoogsteen, K. (1970). *Biochemistry*, **9**, 3976.

Barbet, J., Roques, B. P., Combrisson, S. and Le Pecq, J.-B. (1976). *Biochemistry*, **15**, 2542.

Barbet, J., Roques, B. P. and Le Pecq, J.-B. (1975). *C.r. hebd. Séanc. Acad. Sci., Paris*, **281**, 851.

Berman, H. M., Neidle, S. and Stodola, R. K. (1978). *Proc. natn. Acad. Sci. U.S.A.*, **75**, 828.

Borer, P. N., Kan, L. S. and Ts'o, P. O. P. (1975). *Biochemistry*, **14**, 4847.

Canellakis, E. S., Shaw, Y. H., Hanners, W. E. and Schwartz, R. A. (1976). *Biochim. biophys. Acta*, **418**, 277.

Capelle, N., Barbet, J., Dessen, P., Blanquet, S., Roques, B. P. and Le Pecq, J.-B. (1979). *Biochemistry*, **18**, 3354.

Chan, S. I., Bangerter, B. W. and Peter, H. H. (1966). *Proc. natn. Acad. Sci. U.S.A.*, **55**, 720.

Chan, S. I. and Kreishman, G. P. (1970). *J. Am. chem. Soc.*, **92**, 1102.

Chiao, Y.-C. C. and Krugh, T. R. (1977). *Biochemistry*, **16**, 747.

Chien, M., Grollman, A. P. and Horwitz, S. B. (1977). *Biochemistry*, **16**, 3641.

Cohn, M. and Hughes, T. R. Jr. (1962). *J. biol. Chem.*, **237**, 176.

Davidson, M. W., Griggs, B. G., Lopp, I. G. and Wilson, W. D. (1977). *Biochem. biophys. Acta.*, **479**, 378.

Dervan, P. B. and Becker, M. M. (1978). *J. Am. chem. Soc.*, **100**, 1968.

Dimicoli, J.-L. and Helene, C. (1974). *Biochemistry*, **13**, 714.

Dwek, R. A. (1973). In *Nuclear Resonance in Biochemistry*, Oxford University Press, London, pp. 48, 213.

Dugas, H. (1977). *Acct. Chem. Res.*, **10**, 47.

Emsley, J. W., Feeny, J. and Sutcliffe, L. H. (1965). *High Resolution Nuclear Magnetic Resonance Spectroscopy*, Pergamon Press, New York.

Ezra, F. S., Lee, C.-H., Kondo, N. S., Danyluk, S. S. and Sarma, R. H. (1977). *Biochemistry*, **16**, 1977.

Gabbay, E. J., Adawadkar, P. D., Kapicak, L., Pearce, S. and Wilson, W. D. (1976). *Biochemistry*, **15**, 152.

Gabbay, E. J., Scofield, R. E. and Baxter, C. S. (1973). *J. Am. chem. Soc.*, **95**, 7850.

Giessner-Prettre, C., Pullman, B., Borer, P. N., Kan, L.-S. and Ts'o, P. O. P. (1976). *Biopolymers*, **15**, 2277.

Gorenstein, D. G. (1975). *J. Am. chem. Soc.*, **97**, 898.

Gorenstein, D. G., Findlay, J. B., Momii, R. K., Luxon, B. A. and Kar, D. (1976). *Biochemistry*, **15**, 3796.

Gorenstein, D. G. and Kar, D. (1975). *Biochem. biophys. Res. Commun.*, **65**, 1073.

Gorenstein, D. G. and Luxon, B. A. (1979). *Biochemistry*, **18**, 3796.

Heller, M. J., Tu, A. T. and Maciel, G. E. (1974). *Biochemistry*, **13**, 1623.

Henry, D. W. (1976). In *Cancer Chemotherapy* (ed. A. C. Sàrtorelli), vol. 15, American Chemical Society, Washington.

Hong, S.-J. and Piette, L. H. (1976). *Cancer Res.*, **36**, 1159.

Hollstein, U. (1974). *Chem. Rev.*, **74**, 625.

Jain, S. C., Tsai, C.-C. and Sobell, H. M. (1977). *J. molec. Biol.*, **114**, 317.

James, T. L. (1975). *Nuclear Magnetic Resonance in Biochemistry: Principles and Applications*, Academic Press, New York.

Kallenbach, N. R. and Berman, H. M. (1977). *Q. Rev. Biophys.*, **10**, 137.

Kearns, D. R. (1977). *Ann. Rev. Biophys. Bioeng.*, **6**, 477.

Kondo, N. S. and Danyluk, S. S. (1972). *J. Am. chem. Soc.*, **94**, 5121.

Kreishman, G. P., Chan, S. I. and Bauer, W. (1971). *J. molec. Biol.*, **61**, 45.

Krugh, T. R. (1972). *Proc. natn. Acad. Sci. U.S.A.*, **69**, 1911.

Krugh, T. R. (1974). In *Molecular and Quantum Pharmacology* (eds. E. D. Bergman and B. Pullman), Reidel, Dordrecht, Holland, p. 465.

Krugh, T. R. (1976). In *Spin Labeling: Theory and Applications* (ed. L. J. Berlinger), Academic Press, New York, p. 339.

Krugh, T. R., Hook III, J. W., Lin, S. and Chen, F.-M. (1979). In *Stereodynamics of Molecular Systems* (ed. R. H. Sarma), Pergamon Press, New York, p. 423.

Krugh, T. R., Mooberry, E. S. and Chiao, Y.-C. C. (1977). *Biochemistry*, **16**, 740.

Krugh, T. R. and Neely, J. W. (1973a). *Biochemistry*, **12**, 1775.

Krugh, T. R. and Neely, J. W. (1973b). *Biochemistry*, **12**, 4418.

Krugh, T. R. and Nuss, M. E. (1979). In *Biological Applications of Magnetic Resonance* (ed. R. G. Shulman) Academic Press, New York, p. 113.

Krugh, T. R. and Reinhardt, C. G. (1975). *J. molec. Biol.*, **97**, 133.

Krugh, T. R. and Schafer, W. E. (1975). *J. Magn. Reson.*, **19**, 99.

Krugh, T. R., Wittlin, F. N. and Cramer, S. P. (1975). *Biopolymers*, **14**, 197.

Kuhlman, K. F., Charbeneau, N. J. and Mosher, C. W. (1978). *Nucl. Acids Res.*, **5**, 2629.

Lackner, H. (1975). *Angew. Chem. Intern. Ed.*, **14**, 375.

Lee, C.-H. and Sarma, R. H. (1976a). *Biochemistry*, **15**, 697.

Lee, C.-H. and Sarma, R. H. (1976b). *J. Am. chem. Soc.*, **98**, 3541.

Lee, C.-H. and Tinoco, I., Jr. (1978). *Nature, Lond.*, **274**, 609.

Le Pecq, J.-B. (1971). In *Methods of Biochemical Analysis* (ed. D. Glick), vol. 20, Wiley, New York, p. 41.

Le Pecq, J.-B., Yot, P. and Paoletti, C. (1964). *C. r. hebd. Séanc. Acad. Sci., Paris*, **259**, 1786.

Levy, G. C. and Nelson, G. L. (1972). *Carbon-13 Nuclear Magnetic Resonance for Organic Chemists*, Wiley-Interscience, New York.

Mariam, Y. H. and Wilson, W. D. (1979). *Biochim. Biophys. Res. Commun.*, **88**, 861.

Meienhofer, J. and Atherton, E. (1977). In *Structure-Activity Relationships Among the Semisynthetic Antibiotics* (ed. D. Perlman) Academic Press, New York, p. 427.
Mildvan, A. S. and Cohn, M. (1970). *Adv. Enzymol.*, **33**, 1.
Müller, W. and Crothers, D. M. (1968). *J. molec. Biol.*, **35**, 251.
Patel, D. J. (1974*a*). *Biochemistry*, **13**, 2388.
Patel, D. J. (1974*b*). *Biochemistry*, **13**, 2396.
Patel, D. J. (1976*a*). *Biopolymers*, **13**, 533.
Patel, D. J. (1976*b*). *Biochim. biophys. Acta*, **442**, 98.
Patel, D. J. (1977*a*). *Biopolymers*, **16**, 1635.
Patel, D. J. (1977*b*). *Biopolymers*, **16**, 2739.
Patel, D. J. (1979). *Acct. Chem. Res.*, **12**, 118.
Patel, D. J. and Canuel, L. L. (1976). *Proc. natn. Acad. Sci. U.S.A.*, **73**, 3343.
Patel, D. J. and Canuel, L. L. (1977*a*). *Biopolymers*, **16**, 857.
Patel, D. J. and Canuel, L. L. (1977*b*). *Proc. natn. Acad. Sci. U.S.A.*, **74**, 2624.
Patel, D. J. and Canuel, L. L. (1977*c*). *Proc. natn. Acad. Sci. U.S.A.*, **74**, 5207.
Patel, D. J. and Shen, C. (1978). *Proc. natn. Acad. Sci. U.S.A.*, **75**, 2553.
Piette, L. H. (1974). *Fed. Proc.*, **33**, 1371.
Reed, G. H., Leigh, J. S. and Pearson, J. E. (1971). *J. Chem. Phys.*, **55**, 3311.
Reinhardt, C. G. and Krugh, T. R. (1977). *Biochemistry*, **16**, 2890.
Reinhardt, C. G. and Krugh, T. R. (1978). *Biochemistry*, **17**, 4845.
Reuben, J., Adawadkar, P. and Gabbay, E. J. (1976). *Biophys. Struct. Mech.*, **2**, 13.
Reuben, J., Baker, B. M. and Kallenbach, N. R. (1978). *Biochemistry*, **17**, 2915.
Reuben, J. and Gabbay, E. J. (1975). *Biochemistry*, **14**, 1230.
Sakore, T. D., Jain, S. C., Tsai, C.-C. and Sobell, H. M. (1977). *Proc. natn. Acad. Sci. U.S.A.*, **74**, 188.
Salemink, P. J. M., Swarthof, T. and Hilbers, C. W. (1979). *Biochemistry*, **18**, 3477.
Shulman, R. G. and Sternlicht, H. (1965). *J. molec. Biol.*, **13**, 952.
Shulman, R. G., Sternlicht, H. and Wyluda, B. J. (1965). *J. Chem. Phys.*, **43**, 3116.
Sinha, B., Cysyk, R. L., Millar, D. B. and Chignell, C. F. (1976). *J. Med. Chem.*, **19**, 994.
Sobell, H. M. (1973). *Prog. Nucl. Acid Res. Mol. Biol.*, **13**, 153.
Sobell, H. M. and Jain, S. C. (1972). *J. molec. Biol.*, **68**, 21.
Sobell, H. M., Jain, S. C., Sakore, T. D. and Nordman, C. E. (1971). *Nature, Lond.*, **231**, 200.
Sobell, H. M., Jain, S. C., Sakore, T. D., Reddy, B. S., Bhandary, K. K. and Seshadri (1980). In *Proceedings International Symposium on Biomolecular Structure, Function and Evolution* (ed. R. Srinivasan), Pergamon Press, New York.
Sobell, H. M., Tsai, C.-C., Jain, S. C. and Gilbert, S. G. (1977). *J. molec. Biol.*, **114**, 333.
Stothers, J. B. (1972). *Carbon-13 NMR Spectroscopy*, Academic Press, New York.
Swift, T. J. (1973). In *NMR of Paramagnetic Molecules* (eds. G. N. LaMar, W. Dew Horrocks, Jr. and R. H. Holm), Academic Press, New York, p. 53.
Tsai, C.-C., Jain, S. C. and Sobell, H. M. (1977). *J. molec. Biol.*, **114**, 301.
Ts'o, P. O. P. (1974*a*). *Basic Principles in Nucleic Acid Chemistry*, vol. I, Academic Press, New York.
Ts'o, P. O. P. (1974*b*). *Basic Principles in Nucleic Acid Chemistry*, vol. II, Academic Press, New York.
Wakelin, L. P. G. and Waring, M. J. (1976). *Biochem. J.*, **157**, 721.
Wang, J. C. (1971). *Biochim. biophys. Acta*, **232**, 246.
Waring. M. J. (1965). *J. molec. Biol.*, **13**, 269.
Waring, M. J. (1970). *J. molec. Biol.*, **54**, 247.
Waring, M. J. and Wakelin, L. P. G. (1974). *Nature, Lond.*, **252**, 653.

Wartell, R. M., Larson, J. E. and Wells, R. D. (1974). *J. Biol. Chem.*, **249**, 6719.
Wehrli, F. W. and Wirthlin, T. (1976). *Interpretation of Carbon-13 NMR Spectra*, Heydon and Son, London.
Wells, R. D. (1971). In *Progress in Molecular and Subcellular Biology* (ed. F. E. Hahn), vol. 2, Springer Verlag, New York, p. 21.

Subject Index